THOUGHT YOU'D ENJOY
LOVE DAVE :)

CW01510390

British Naval Armaments

Edited by Robert D Smith

ROYAL ARMOURIES
Conference Proceedings 1

Published in 1989
by the Trustees of the
Royal Armouries
HM Tower of London
London EC3N 4AB

Designed by Tina Lawlor

©The Trustees of the
 Royal Armouries, 1989

ISBN 0–948092–11–4

Printed by Henry Ling Ltd., The Dorset Press, Dorchester, Dorset

Contents

Foreword by the Master of the Armouries

This is the first of a new series of proceedings devoted to publishing papers given at conferences held at the Royal Armouries. Recently the Royal Armouries has hosted two conferences on recent research into the history and development of artillery. The first of these, *Guns from the Sea*, held in 1986, was published by the Nautical Archaeological Society in 1988. The second, held in 1987, is the subject of this volume and traces the development of artillery from the 17th to the end of the 19th century.

This period saw many changes in artillery, Prince Rupert's attempts to improve cast-iron cannon, the evolution of various patterns by the Ordnance Office, the development of the carronade and the enormous technological advances which took place during the last 50 years of the 19th century. The history of artillery in the 18th century is inextricably linked with the progress of what we call the Industrial Revolution. In addition warfare at sea changed beyond recognition because of the ever more powerful armaments with which ships were equipped.

The conference which is the subject of this first volume of proceedings attracted an international audience and was addressed by speakers from several countries. However its success was largely due to the efforts of the staff of the Royal Armouries who worked long hours making all the necessary arrangements while at the same time preparing their own academic contributions, some of which are published here. All the contributors to the conference and to this volume owe their gratitude to Robert Smith, on whose shoulders much of the burden of organisation lay and who subsequently edited the contributions which appear here.

Further volumes in this series will appear in the future as we develop our programme of conferences over the coming years. I look forward to introducing the proceedings of other artillery conferences and to involvement in an increasingly wide range of conferences in the Royal Armouries.

G M Wilson
October 1989

Introduction

ROBERT D SMITH AND RUTH R BROWN

The papers which make up this volume were presented at the conference *British Naval Armament 1600–1900*, organised by the Royal Armouries in November 1987. This conference was held as a follow-on to one held the previous year entitled *Guns from the Sea*, which had dealt mainly with the period up to 1600.[1]

The idea for a meeting to discuss the subject of ordnance originated from a meeting in 1986, where it was realised that a forum was needed to put forward current research ideas and projects and to bring together those working on the history of artillery development. Due to the extensive material available and the wide range of subjects and topics to be covered, two conferences, divisable by date, seemed the obvious solution, 1600 being the dividing line. After the success of the 1986 conference, we decided to extend the duration of that planned for 1987 to two days. The first day, at the Tower of London, included a broad coverage of the period as well as papers on various aspects of gunnery, gunfounders and on particular guns. The second day was held at Fort Nelson, one of the Palmeston Forts overlooking Portsmouth and now a Royal Armouries outstation, housing its major collection of artillery. This gave the participants an opportunity to tour the fort as well as to attend lectures.

The more recent history of artillery is, for the most part, relatively well-documented, especially when compared to the period prior to 1600. However, there are still large areas which are poorly researched or about which less is known. The conference was seen as a means whereby some of these areas could be discussed. It included papers on individual guns and broad surveys, such as on the development of naval guns in the second half of the 19th century. This latter paper, by John Campbell, could not be presented at the conference, but is included here because it provides a thorough summary of the developments from smooth bore naval guns to the complex built-up guns developed by 1900.

Several papers presented at the conference are not included or are only summarised here for a variety of reasons. Sarah Barter-Bailey talked about the guns invented by Prince Rupert and produced by the Brownes in the late 17th century. Ms Barter-Bailey is currently working on a comprehensive study of this subject to be published by the Royal Armouries. Graeme Rimer presented a paper on the armament of *HMS Inflexible* and the Dover Turret, which will also be published at a later date as part of a wider survey. Martin Dean gave a personal, and at times deeply pessimistic, view of the current state of nautical archaeology. Peter Jones gave a paper outlining the terminal ballistics of artillery projectiles.

A conference such as this inevitably relies on a great many people to ensure that all goes smoothly. As ever the Trustees of the Royal Armouries and Guy Wilson, the Master of the Armouries, gave their full support. We must thank all the speakers for making the proceedings both informative and enjoyable. We must also thank many of the staff of the Royal Armouries who helped with organisation both prior to and during the conference. In particular we must thank, Denise Ferry who typed the preprints, Gary Kennard who manned the projector, Jeremy Hall who produced many of the photographs, and Tina Lawlor who saw the papers through the printers. At Fort Nelson delegates were treated to a display of artillery drill by the Portsdown Artillery Volunteers — Ian Maine, Ian Mountifield, Duncan Williams, Geoffrey Salter and Graham Walters. Nicholas Hall, Frank Strugnall and David Moore acted as guides at the fort and showed delegates around. We also owe many thanks to Helen Strugnall, Mavis Strugnall and Celia Salter who managed to produce a quite wonderful lunch at Fort Nelson, in conditions which were primitive in the extreme!

Notes

1 *International Journal of Nautical Archaeology*, Vol. 17, Number 1, 1988.

Guns and Ships

D J LYON

The use of guns at sea had a progressive impact on ship design from the late middle ages to the end of the 19th century.

Initially, guns were placed on the upper-works, then later within the hulls, of existing ship types (carracks, hulks etc). The standard Mediterranean fighting ship, the galley, was also modified by placing a fixed gun or guns in the bows. By the mid-16th century, purpose-built gun-armed warships were appearing, equipped for all-round fire, for example the *Mary Rose*. The specialised small-oared 'gunboat' designed for one or two large guns, also appeared in the form of the 'roo-barge', a type which was revived in the late 18th century.

By this time a distinction between the large fighting ships and small fast vessels, for scouting and other duties, was emerging. However, it was not until the first Anglo-Dutch war that these developments reached their logical conclusion — concentration on broadside fire, the tactic of fighting in line and the development of the ship of the line.

Towards the end of the 17th century, the first real weapons system appeared, with a ship designed round her weapons. This took the form of the bomb vessel, a French invention. Later, British versions were equipped with turntable-mounted mortars — probably the origins of the turret. This precedent was then followed by the turntable-mounted long guns of the innovative in-shore craft, designed by the Swede, Chapman.

The frigate, evolved by the French in the second quarter of the 18th century, with its 'tween deck, marked a new way of mounting guns in the hull of an ordinary ship. The early 19th century saw larger guns with increasing use of centre-line mountings, pivots and racers for traverse.

Guns were now becoming integral to both hulls and design. This process accelerated with the development of iron and later steel hulls, steam power and the race between guns and armour. The box battery and even the true turret were only temporary solutions. By the end of the 19th century warships were designed round integral structures of magazine, barbette and gunhouse. Their size gradually increased due to the secondary armament needed to combat the torpedo and the craft that carried it — the torpedo boat. This innovation was quickly followed by first the submarine and then aircraft.

Evidence for the Use of Cartridges in Artillery 1560–1660

S BULL

Despite the amount of interest in early modern artillery, particularly that aroused by nautical archaeology, there are many areas of gunnery technique which have yet to receive detailed examination. Loading is perhaps the most vital of these. Several authorities have assumed that the process of loading was necessarily slow, usually being carried out with the use of the ladle.[1]

The ladle was given some prominence in contemporary literature. At its simplest it was like an elongated grocer's scoop, constructed of sheet copper or an alloy, bent round a former and fitted to a stave. The use of non-ferrous metals minimised the possibility of a spark being struck during loading. The capacity of the ladle was usually calculated in terms of volume, using the shot diameter as a unit of measure. If the ladle were not to be uncomfortably large, the load was divided by two and the ladle used twice to charge the gun. This technique meant that a powder barrel had to be kept close at hand during firing. The use of a 'budge' barrel, with a leather cover and drawstring, or other method of closure, helped to prevent fire reaching its contents (Figure 1).[2]

During the late sixteenth and early seventeenth centuries many ladles were supplied to forces by the Ordnance Office at the Tower. Very often materials such as copper plate were furnished by merchants, for example Edward Fawconer, Richard Cockyn (or Cockaine), Robert Evelyn or Nicholas Blaque, and the pieces were finished and staved by Ordnance employees, contractors or London pikemakers (Figure 2).[3]

The many difficulties of loading with the ladle are readily apparent. In the heat of battle, the measures obtained are not likely to have been any more exact than numbers of heaped spoonfuls. Henry Hexam helpfully suggested *a little jog (to the ladle) that the surplus may fall down again into the barrell.* Thomas Smith favoured a brisk tap with a ruler.[4] Another drawback was the slowness of the method. Using the ladle two or more times took longer and, of course, a whole barrel of powder had to be moved about. Lastly, and most importantly, ladle loading was highly dangerous. Loose powder could fall to the ground or onto a wooden deck and be ignited by flash or by embers, dust could rise from the powder, and additionally, the gunner's burning match was a danger both to the ladle and to the barrel of powder. The matchlocks of the infantry and enemy action could double the problem. Walking back and forth with an open powder barrel cannot have been pleasant at the best of times, but in action it was tantamount to suicide. On a ship it is very difficult to see how this method could have worked at all.

William Bourne summed up the difficulties of the ladle in his *Shooting in great ordnance.*

> *The ladell shall have sometyme more pouder, and sometyme less pouder, by a good quantitye, and especially if that hee dothe it hastely as in time of service it always requireth haste.*

Spilt powder could result in the *spoyling of men* and, in short, there was *no worse lading or charging of Ordnance than with a ladell.*[5]

It would therefore seem obvious that the use of cartridges — sewn or glued bags of paper, linen or canvas — was a highly practical

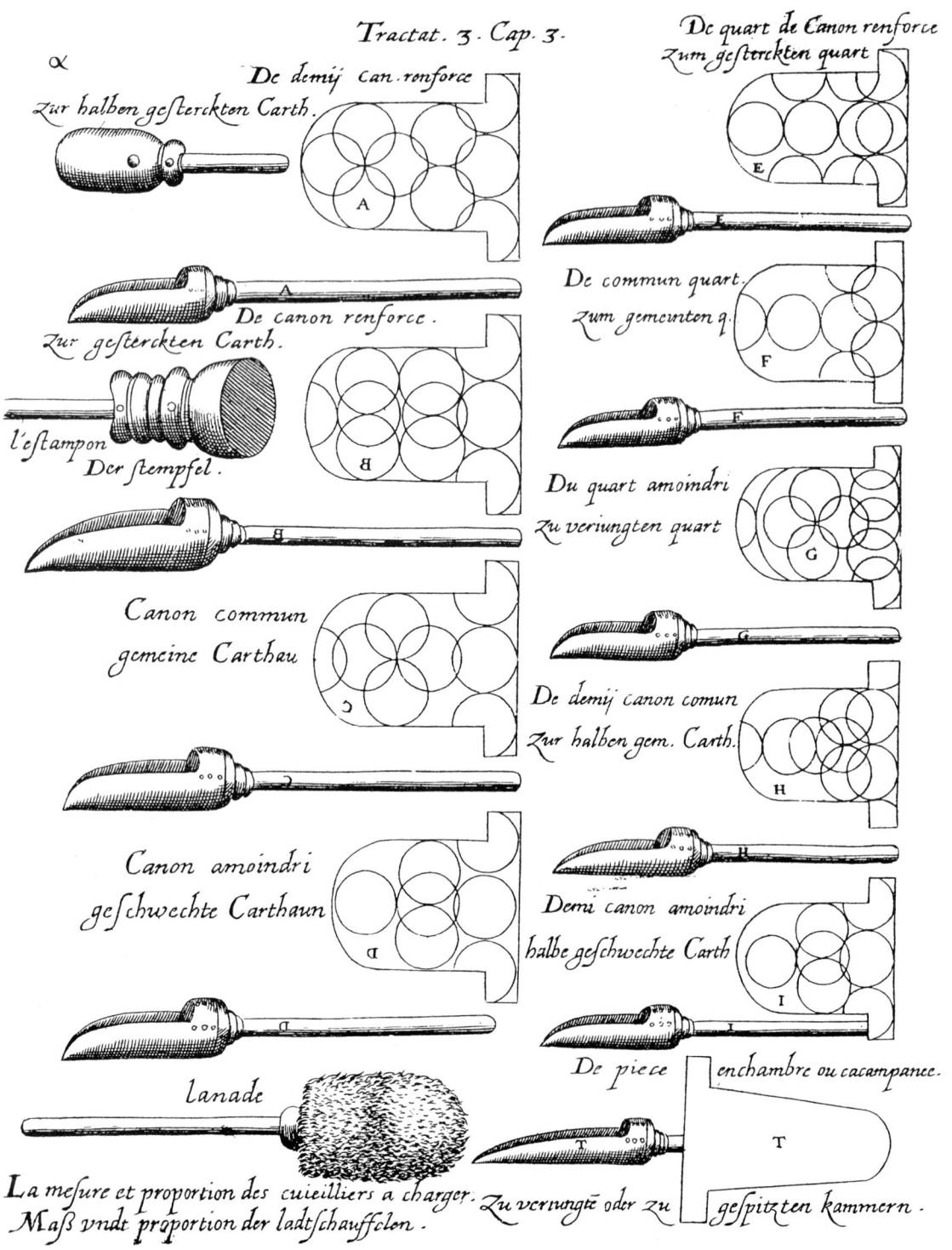

Figure 1   Ladles, rammers etc in Norton *The Gunner*.

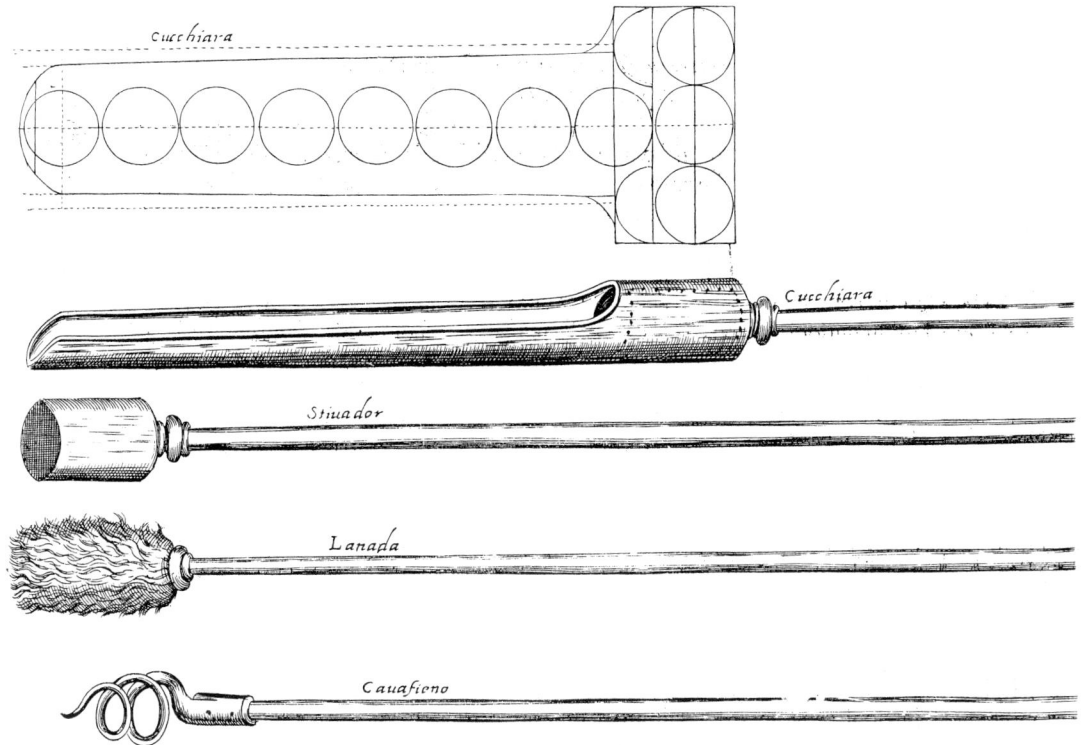

Figure 2   Ladle, rammer, sponge and worm in Pietro Sardi *L'Artiglieria*, Venice, 1621.

solution (Figure 3). Contemporary manuals seem remarkably consistent on this point. As Peter Whitehorne observed:

> *For the more speedie shooting of ordinance, the iuste charge in pouder of everye peece may aforhand be prepared in a readinesse and put in bagges of linnen or in great papers made for the same purpose, which in a sodaine may be chapt into the mouth of a peece with the bollet thereof thrust after, as farre as they will goe, and then thrusting a long wyer into the touchehole with some pouder so soone as it is leveled, it may incontinent be shot of: which maner of charging is done most quikely and a great deale sooner than any other waye, and when haste requires very needfull.*[6]

This sort of advice was repeated frequently during the next century and most authorities agreed that the cartridge was the best method of loading, particularly in action.

Every manual gave instructions for the making of cartridges. Amongst the most important English examples were Bourne's *Art of shooting in great ordnance*, 1587, Lucar's version of *Tartaglia*, 1558, Thomas Smith's *Art of gunnery*, 1600 and William Eldred's *Gunners glasse*, 1646 (Figure 4). Nathaniel Nye in his *Art of gunnery* described the making of cartridges in the following terms:

> *take canvas, such as the powder will not creep thorow, and let it be in breadth . . . just three diameters of the peece . . . and for the length you will find it by the filling of them, these being sewed together upon a mould; which must be a very little lesse than the diameter of the bore, and about 4 diameters long.*[7]

Paper cartridges could be made in a similar fashion, but in this case the former was first smeared with tallow to prevent sticking and the seams were glued.[8]

5

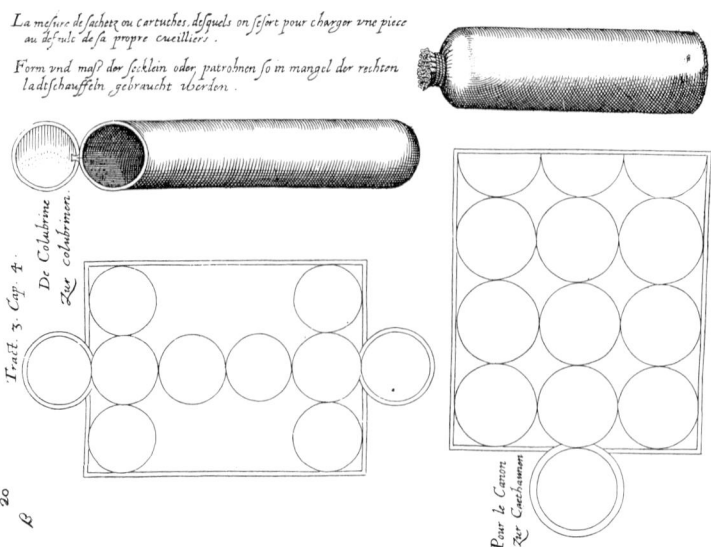

Figure 3   Cartridge construction as pictured in Robert Norton *The Gunner*, London, 1628.

Parallel examples for cartridge manufacture can be found in every major European language. Diego Ufano's celebrated *Trato de Artilleria* clearly illustrates the point. The original edition, in Spanish, was published in Brussels in 1613. Here the reader was informed of the making of 'el cartucho' and how handy a way this was to charge ordnance. In the French edition of 1621, the cartridge translates to 'patron' and 'sachets'. In the German editions these became 'Secklein oder patronen'. In Norton's *The gunner*, the diagrams by De Bry for the continental editions of Ufano were pirated and put straight into an English text — noted as 'cartridges'.[9] The Italians similarly were not left out as many of their texts also mention cartridges. The best example was perhaps Luigi Collado's *Pratica manuale*.[10]

All of this goes to show that the making of cartridges was common knowledge. Proof that they were actually used on a regular basis is more difficult to establish, although some archaeological evidence, such as the presence of reamers aboard the *Mary Rose* makes this seem likely.[11] It is noticeable, however, that cartridges were not a central issue store. In England, prior to the Civil War, it was not usual to find cartridges sent from the Ordnance Office at the Tower to ships and garrisons in the same way as guns and barrels of powder. Furthermore, there is evidence to suppose that it was the duty of individual gunners to make up their own cartridges, as Eldred says *at spare times . . . in garrisons or other places.*

This was similarly true of ship's gunners, as is suggested by Richard Hawkins in his *Observations*, where he relates the loss of his ship *Dainty* in action with the Spaniards. He blames the master gunner,

> *For bearing me ever in hand, that he had five hundred cartreges in a readinesse, within one hours fight we were forced to occupie three persons, only in making and filling cartreges, and of five hundreth Elles of canvas and other cloth given him for that purpose, at sundry times, not one yard was to be found.*[12]

This may be exaggerated, but clearly cartridges were the norm and it was the gunners' duty to make them.

The best evidence that cartridges were used as a matter of course is in the supply of materials for their construction, and in

A fourme for a Cartredge.

Figure 4 Cartridge construction showing a 'former' from Cyprian Lucar's appendix to *Tartaglia*, London, 1588.

'remains', or lists of stores returned after voyages. We have many excellent examples of the provision of cartridge-making materials to the ships of the Elizabethan navy. When the *Hoape* was fitted out in 1572, cartridge formers featured in its stores. In 1572 the lists for the *Swallow* mentioned not only formers, but *canvas for cartouches xx elles* and three reams of *royal paper*. In 1597 Richard Ascue, purser of the *Warspite*, delivered 57 yards of canvas to the master gunner William Bull at a cost of 71s 3d. We can trace many similar deliveries, explicitly for cartridge-making purposes.[13] In the *Book of the Remaynes* taken in 1595 and 1596 for returning Royal Navy vessels, every ship has been provided with cartridge-making materials, most of which had been expended by the time the fleet returned.[14]

The records are yet more specific for the early Stuart navy. The fleet which went to Cadiz in 1625, set out with formers, canvas, Paper Royal, glue and thread amongst the gunnery stores. When the ship returned the clerks of the Ordnance were able to enumerate the finished cartridges aboard. In the fleet of 1639, the minutely detailed account shows that the cartridges were made up early in the voyage, and that much powder was wasted in the process especially if the cartridges were emptied again at a later stage. However, it is possible that a certain percentage was accepted as a perk of the job.[15]

We have similar information on land garrisons, if not so complete or organised. Cartridge-making equipment was sometimes supplied and gunners sometimes petitioned for more. This was the case in 1629, when one of the gunners at Dover wrote to Lord Zouch, Warden of the Cinque ports, requesting a

multitude of supplies including — *Royall Paper to make carthridges withall*.[16]

Even if the Ordnance Office did not actually supply the finished cartridge, it certainly provided materials and the inventories suggest there was usually quite a lot in store. In 1559 stocktaking revealed 125 years of canvas for cartridges and ten reams of Paper Royal. The 1635 list shows not only materials, but over 600 formers. Every inventory of the sixteenth and seventeenth century shows some provision for cartridge manufacture.[17]

Evidence relating to field armies also reveals plenty of cartridge making materials. In 1620–1 an expedition was planned to go to the aid of the Elector Palatine. Equipment included *canvas for cartouches 1000 elles at* 6d *the ell* — no less than three quarters of a mile of cartridge-making canvas. In 1627 an expedition to the Isle de Rhe took with it at least five hundred yards of canvas and two tons of cartridge paper. We should not imagine that these stores were intended for any other purposes, for there were plentiful supplies of writing paper and canvas sandbags over and above the provisions for cartridges.[18] During the Civil War cartridges were carried along with the field guns. The Scottish gunner and theorist Thomas Binning suggested a universal scale of issue of 24 cartridges per gun, of which half were to be filled and ready for use at all times. A number of entries in Roys *Royalist Ordnance Papers* suggest this was the sort of high standard which they wanted to attain.[19]

The case of the New Model Army is particularly interesting, as this body was provided not only with cartridge-making materials but, unusually, with finished cartridges as well as with cartridges cut out of cloth but not sewn. This may give us a date at which cartridges began to be considered an issue item, rather than something which it was the duty of individual gunners to make up. No doubt this was made possible by the greater standardisation of armament in the New Model Army, in which field pieces of 3 lb and 6 lb were normal. In earlier times, when guns were much more individually styled, it was not worthwhile considering the cartridge as an issue store, because so many different types were required. One of the largest orders for finished cartridges was contracted on 10 January 1646 with Nathaniel Humfreys and Richard Bradley for 1,000 *at ten pence a peece . . . ready money*. It was highly unusual that ready money be either asked for, or given, on a government contract and this may be a measure of the importance placed on the supply of cartridges.[20]

Such strong evidence can only suggest that the cartridge was the normal method of loading used from the early sixteenth century onwards. The ladle in the manuals and in archaeological finds requires some reinterpretation. Two possible explanations may reasonably be put forward. First, it was an added security. If the cartridges were insufficient in number, or damp, or lost, the gunner could have resorted to the ladle for loose powder as an emergency measure. It is even possible that the ladle could have been used for the quick insertion of the cartridge. Second, in ceremonial or practice firing, speed was not important. It was the quality of the show that mattered. In such circumstances few gunners would have bothered with the expense of a cartridge. We have therefore the highly-choreographed descriptions of the use of the ladle — handled as Eldred said like *an artist*.

Notes

1 See for example O F G Hogg, *English Artillery*, Woolwich, 1963, p. 45; H W L Hime, *Gunpowder and Ammunition*, London, 1904, p. 235.
2 See R Norton, *The Gunner*, London, 1628, pp. 42–3, and folding plates.
3 Public Record Office, War Office papers 54/3; 6; 7; 8; 9; 11.
4 H Hexam, *The Principles of the Art Militarie*, The Hague, 1640, Part III, pl. 3; Thomas Smith, *The Art of Gunnery*, London, 1600, p. 81.
5 W Bourne, *The Art of Shooting in Great Ordnance*, London, 1587, pp. 30–1.
6 P Whitehorne, *Certain Wayes for the Ordering of Souldiers in Battelray*, London, 1573, p. 33v.
7 W Nye, *The Art of Gunnery*, London, 1647, p. 42.
8 See J Roberts, *Compleate Cannoniere*, London, 1639, p. 29.
9 D Ufano, *Trato de Artilleria*, Brussels, 1613, p. 306; *Artillerie C'est a Dire*, Zutphen, 1621; *Archeley*, Frankfurt, 1614.
10 L Collado, *Pratica Manuale Di Arteglieria*, Venice, 1586.
11 The 'reamer' or priming iron was a sharp spike or wire to clear the touch hole and pierce the skin of the cartridge.
12 *The Observations of Sir Richard Hawkins Knight, in his Voiage to the South Sea Anno Dom 1593*, London,

1622, p. 127. An 'Ell' was a unit of measure equalling 1.25 yards in English usage. (A Scottish 'Ell' was 37.2 in; the Flemish 'Ell' 27 in.)
13 British Library Additional MS 5752 f36; PRO WO 54/2. 'Paper Royal' probably ment initially any large sheet. In the printing trade it was later 25 in. by 20 in. The term is known to have been in use from the late fifteenth century.
14 National Maritime Museum MS CAD C/1; ADL/H/14; PLA/P11; PRO WO 55/1627.
15 PRO WO 49/110; 55/1601.
16 British Library Egerton MS 2584, f362.
17 PRO WO 55/1672; 55/1690; State Papers 12/6 etc. See also H L Blackmore, *The Armouries of the Tower of London*. Vol 1 Ordnance, London, 1976, pp. 251–389.
18 British Library Harleian MS 5109. Royal Artillery Institution MD 979 'Inventory of the Equipment of the Artillery Embarked', 1627.
19 I Roy, *The Royalist Ordnance Papers*, Oxfordshire Record Society, 2 Vols 1964 and 1976, passim. T Binning, *Light to the Art of Gunnery*, Edinburgh 1676, p. 109.
20 Museum of London MS 46–78/709. See also G I Mungeam, 'Contracts for the Supply of Equipment to the New Model' *Journal of the Arms and Armour Society*, Vol VI, part 3, 1967.

John Browne and Prince Rupert's guns

S BARTER BAILEY

A patent for *preparing and softening all cast or melted iron so that it may be fyled and wrought as forged Iron is* was granted to Prince Rupert in 1671 for a term of 14 years. Although the patent did not mention any military application and did not, in fact, describe the methods to be used, the only recorded use of the technique seems to have been in the manufacture of military and naval stores, specifically cannon, grenades and anchors.

The first examples seem to have been small cast-iron three-pounder guns for a new royal yacht, the *Cleaveland*, which was cast by one of the normal suppliers to the Ordnance Office, and then treated in a furnace and bored in a mill set up at Eton and Windsor, where Prince Rupert was Constable of the Castle. These were received into store in December 1671 and seem to have been successful enough to encourage the Ordnance office to support further experiments. These took place at Windsor, in the Office's own establishment at Woolwich and by the Gunfounder to the Office, John Browne, during 1672 and early 1673.

Of these, the only ones who eventually produced patent guns in any number were, not surprisingly, those who had access to large-scale industrial resources — the Weald-based Browne gunfounding partnership. Their first large delivery was made in June 1673 and between then and September 1677 regular deliveries to the final value of £52,995 were made. After September 1677, however, a disagreement seems to have developed over the cost and, perhaps, over the ultimate value of the patent guns and further deliveries were only accepted grudgingly and at much reduced prices. There is no evidence that any more were manufactured after the late 1670s and, although some were delivered as late as the 1690s, they seem to have been guns produced in anticipation of further orders in the 1670s and stored until needed.

The technique patented seems to have been a method of annealing the guns in large furnaces and then turning them, both to polish the bore and to smooth out any flaws in the casting. They were known throughout their life as 'turned and nealed guns' and claims were constantly made that they were as accurate and as handsome as bronze guns. It does not seem that the basic method of casting cannon was affected, indeed there are references to treating existing guns. They were normally issued as sea service or garrison guns, or used in such 'second-rate' areas as the 'train of artillery' sent to Ireland, with the exception of large numbers of 3-pounders cast for yachts of all types. Such evidence as has been found, indicates that accurate machining was the aim that Prince Rupert was pursuing and there is some evidence that it was achieved, although not sufficiently to justify the much greater cost of the treated guns as compared to the cost of 'rough iron guns'. When the British government lost interest in them, attempts were made to export them to France and some were actually sold to the Venetian Republic. Only three surviving examples of these guns have been identified, in the Museum of Artillery at Woolwich, and one of them is illustrated in Figure 1.

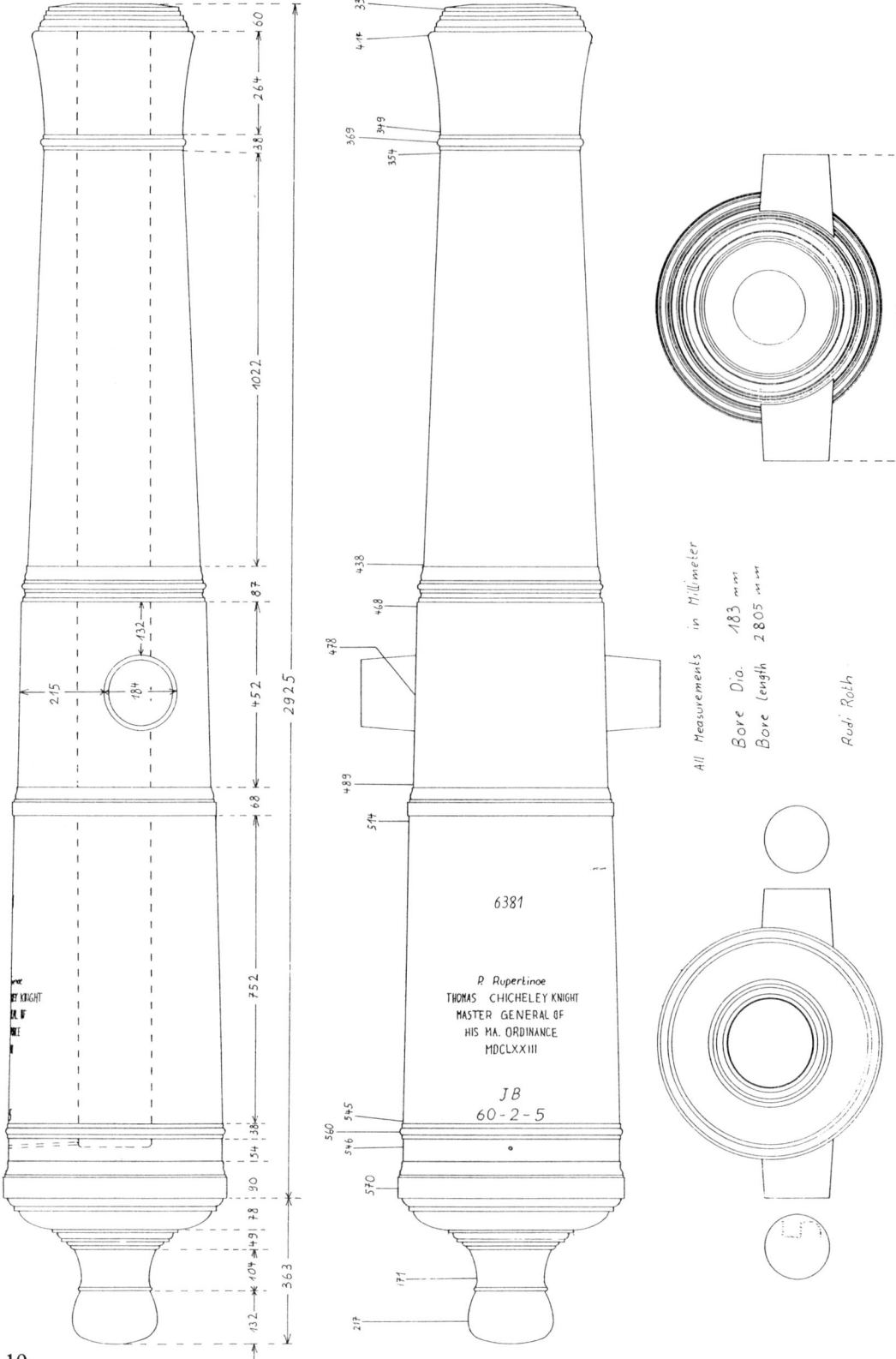

All Measurements in Millimeter

Bore Dia. 183 mm
Bore Length 2805 mm

Rud: Roth

R. Rupertinoe
THOMAS CHICHELEY KNIGHT
MASTER GENERAL OF
HIS MA. ORDINANCE
MDCLXXIII

6381

JB
60-2-5

Figure 1   Gun cast to Prince Ruperts Patent probably by John Brown (III 13 Museum of Artillery Woolwich drawn by R Roth).

10

British Artillery Design

A B CARUANA

The word 'design' can be defined as a plan, a scheme or way of doing something. We can take this definition a stage further if we apply it to the gun — the design of a gun to achieve a particular end. In a piece of artillery, this end is invariably to aim a projectile in a particular direction and in such a manner so as to hit the target.

The aim of this paper is to describe the chronological sequence of the various designs of the gun.

A gun is basically a tube closed at one end and open at the other. The method of loading, whether by the breech or the muzzle, is immaterial. At the moment of discharge both types are the same, closed at one end and open at the other.

Artillery poses particular design problems. The main problem is to contain the pressures produced on the fabric of the gun by the burning of the propellant. At the same time, the gun should be as light as possible to enable movement and handling. Bad design produces either a gun that is unnecessarily heavy, or a gun incapable of resisting the pressures produced by the propellant. These pressures will vary according to the changes in the propellant or the resistance offered by the projectile. For example, a gun which is capable of withstanding the pressure of the propellant in use at one time may not be able to do so if there is a major improvement in the quality of the power of the propellant. Gun design is a constant battle between the designer who seeks to contain the pressure and the powder manufacturer who seeks to increase it.

I do not propose to discuss the early Tudor designs of the cast gun, or howitzers or mortars. By 1600 the cast gun was predomi-nant, both in brass and iron. Brass guns were usually used at sea and for mobile operations and iron guns in static positions. However, there were exceptions. There were both light and heavy brass guns. Double reinforced gun, with a muzzle swell for sea service and muzzle rings for land service were usually produced. Older types were certainly still in service and many obsolescent guns were used until they were no longer serviceable. At any one time there was a great variety of types in use. The only gun which it is possible to date with any accuracy is the gun in production at any one time, and the only guns recorded are government pieces.

The designers of the guns produced in 1600 are unknown. The men who cast them are not, but the subject of gunfounding is beyond the scope of this paper. Provided the founder cast the gun honestly his name is immaterial and, if he did not, the piece would fail at proof. What is certain at this time, is that gun design was centrally controlled by the Board of Ordnance, the particular official concerned being the Surveyor of the Ordnance.

At this stage guns were designed on the principle of 10 diameters of the bore being equal to the external circumference at the touch hole, although light brass guns were considerably less. This general principle lasted until the English Civil War. Few guns were produced during the reign of Elizabeth I, and fewer during the reigns of James I and Charles I, who used many Spanish guns. Whether this was a direct result of the Armada, or an agreed traffic between the two countries, is uncertain. However, immediately prior to the Civil War, a light brass gun was produced. This was cast between 1638 and 1640 by John Browne. There are two examples in the Royal Armouries, described by Blackmore as 4 pdrs, but which are in fact 'brass drakes of 3 lb bullets'. Two types were cast; the heavier averaged 2 cwt, 3 qtrs (2–3–0), and the lighter 2–0–20. They were about half the design size of the heavy brass gun, being only $5\frac{1}{4}$ diameters in circumference at the touch hole. Such a gun would not have withstood the pressures produced by firing round shot, and was probably designed for case shot. This is not the earliest record of this type as it is recorded at the Siege of La Rochelle in 1627.

The first major improvement in ordnance design occurred after the second Dutch War. The celebrated Dutch Raid of 1667 provoked a national panic and, subsequently, a rearmament programme. This resulted in the Rupertinoe guns, produced by George Browne, the gunfounder, according to a patented process developed by Prince Rupert.

Eighty years later this gun was still being cited as an example of good design and was probably the best designed smooth-bore gun ever produced in Britain, certainly considerably better than those produced during the 18th and 19th centuries. They were designed on the basis of eleven diameters at the touch hole diminishing to seven diameters at the neck.

The next apparent innovation was Bellingham's gun. This was a Falcon (bore $2\frac{3}{4}$ inch) made in two pieces, which probably screwed together. There are only sixty recorded, paid for on 1 January 1690/1. The idea was promptly taken on board by the Surveyor of the Ordnance, who had the Wightmans cast a Falconet ($2\frac{1}{4}$ inch) (which definitely did screw together), in mid-August 1691. Thereafter the idea seems to have been dropped. It is, however, worthy of note as it is the first record of the screw-gun concept.

During this period gun design gradually degenerated, which is the central tragedy in the history of British ordnance design. This was probably due to political factors. The knowledge of sound design had been gradually improved, partly by experience and partly by

scientific experiment, since the reign of Henry VIII. The Commonwealth period can be regarded as a temporary hiatus, since this knowledge was not lost. However, after the Glorious Revolution of 1689, all Stuart office-holders were regarded with reservation at the very least and Lord Dartmouth, the Master General of the Ordnance, and Henry Sheers, the Surveyor, both ended up in the Tower. William III brought with him his own adherents, led by the Duke of Schomberg, who became Master General of the Ordnance. He was killed at the Battle of the Boyne. However, the Board of Ordnance was thoroughly purged and most design knowledge therefore dispersed and lost. Exactly the same happened in 1714, when Queen Anne died and George I succeeded to the throne, when even the last remnants of the knowledge of sound design were lost. British artillery design never recovered from this position since our designers made the classic error, thenceforward, of copying French design.

The first Georgian designer was Albert Borgard. He was an eminent soldier with an excellent practical knowledge, but a limited understanding of design theory, and a natural inclination to continental rather than English theory, since he was, by birth, a Dane. Borgard was the first and last man to design a complete system of artillery. He disposed of the nominal terminology in guns and he probably introduced the Royal calibre. His howitzers and mortars were the models used for design for a century. With guns he was, however, less successful. His small iron calibres were sound, but his middle-sized calibres failed. His large iron guns were never cast, which is ironic, since they were well designed guns. Of his brass guns, only the $1\frac{1}{2}$ pdr remained in production for any significant length of time.

Borgard's designs were accepted in 1716 but he was supplanted as the official designer in 1722 by John Armstrong. This was the result of patronage. When Borgard's patron, the Duke of Marlborough, died in 1722 he had no protector, and Armstrong had influential friends. If Borgard had his limitations as a designer, Armstrong was a disaster. Armstrong had always been employed as an engineer and all his artillery designs failed. He started by redesigning the Borgard system in 1722 and 1724. He brought out his own design in 1727, heavily French influenced, and redesigned it in 1732.

Armstrong's 1732 regulation lasted no longer than he did, both died in 1742. There was a complete redesign in 1743 and 1744, a partial redesign in 1753, and finally, the regulation of 1764, which was the result of proposals put forward by Charles Frederick, the Surveyor of the Ordnance, in 1760. The major change was a reduction of windage, the diameter difference between shot and bore. Such a reduction increased range, accuracy, weight, recoil, and chamber pressures, all aspects of performance. These guns were reasonably sound, but their basic design was bad, a deficiency compensated for by their weight, thus making them disproportionately heavy. It was with these guns that the British fought the American War of Independence. It is these guns, with their G3 cypher, that survive in large numbers today. Because the form of the Armstrong pattern was largely retained, these guns are generally known as Armstrong guns but they are really Fredericks guns. However, in my opinion, the most appropriate term is Armstrong-Frederick.

In addition to bad design was the problem of new methods of production. For iron guns, production was centred on the Weald, an area of south-eastern England, where there was iron ore, water, for power and transport, and wood for fuel. The loam of the area was particularly good for moulding. Wealden iron guns were cast at comparatively low temperatures and the molten iron flowed freely and shrank little when cooling. The guns were cooled slowly, resulting in fewer stresses in the metal. However, in the mid-eighteenth century production moved to the Midlands,

Scotland, and, to a limited degree, to South Wales. Sand and flask moulding were introduced, and with them a much faster cooling cycle. In addition, the Midland and northern ores were harder and needed higher temperatures for smelting. Coke was introduced to achieve this, which resulted in a higher sulphur content. A different metal was produced, harder, more brittle, and more stressed. The different types of ore also produced their own characteristics.

As a result, when Thomas Blomefield became Inspector of Artillery, in 1780, and proved guns more strictly, he rejected nearly half of them. Between 1782 and 1785 his department carried out a general reproof of ordnance. In 1787 his own design of iron guns was cast. This design was very similar to that of General Jacques Charles de Manson in France, so similar that it is in many ways a copy. Blomefield eliminated the prominent mouldings of the Armstrong type, a source of weakness, made the first reinforce almost cylindrical and gave his guns a strongly tapered second reinforce and a strengthened chase. The result was a heavy but sound gun. It represented an improvement in design, but did not solve the fundamental problems

By far the best designer of guns was Thomas Desaguliers, whose output was small but significant. He designed only brass guns, which were long and slender, but which shot well and were very reliable. Desaguliers died in 1780, by which time he had produced the first horse artillery carriage and gun. His field guns continued in use, in spite of transport difficulties, until well into the 19th century.

William Millar introduced new gun designs in the 1820s, but they represented no improvement in design. His naval shell guns were really heavy howitzers, and his 32 pdr gun — everyone designed at least one 32 pdr — represented no advance in design.

In 1838 Mr Monk, the chief clerk of the Royal Gun factory, introduced a new pattern. He designed only the heavy calibres — 32, 42 and 56 pdrs. Monk's design is remarkable in two ways: he was the first civilian gun designer and the worst of all designers. His design added both a plane of weakness in the area of greatest pressure and increased the thickness of metal beyond the point at which it had any effect. He understood neither the purpose for which he was designing, nor the material he was proposing to use.

The last major designer was William Dundas. Fortunately his design was better than Monk's, but was restricted to the external pattern introduced by Millar, so that his guns still had the fatal flaw of the plane of weakness. His major achievement was the 68 pdr of 95 cwt, first cast in 1847.

The final test of smooth-bore ordnance was the Crimean War. It did not totally fail, but it spectacularly exposed all its weaknesses. The Lancaster gun experiment failed, iron guns blew up and heavy sea service mortars split down the middle. The result was the 1857 designs of Francis Eardley Wilmot. These incorporated the knowledge of planes of weakness, published by Robert Mallet in 1856, and finally disposed of external ornament. It was previously thought that his designs were not cast, due to the introduction of the compound gun in 1859, but his mortars certainly were and it seems possible that his howitzers may have been. Eardley Wilmot was the last SBML designer and his designs were sound.

Carronades and Blomefield Guns

Developments in Naval Ordnance, 1778–1805

B LAVERY

The carronade and the Blomefield pattern gun were the fundamental armament of the British navy for most of the French Revolution and throughout the Napoleonic wars. The Blomefield gun also equipped fortresses, and dozens of examples are still to be seen throughout Britain and its former colonies. The carronade was used rather less for land service, but it was copied extensively in other countries, notably France and the USA.

Of the two types of gun, the carronade was of course the most radical, and since it was the first to be produced, it is considered first. This paper is not intended to be a complete history of the carronade but will attempt to draw attention to certain aspects of the subject which until now have been neglected.

The carronade was a radical design, but the basic idea was not new. Earlier guns of various types are often described as 'forerunners of the carronade'. In fact it is more useful to consider the carronade merely as the final episode in a recurrent pattern of gun development. Whenever naval tacticians favoured close-range action, gun-founders began to make weapons which would achieve this. If we look at the seventeenth century alone, we find that first the 'drake' was developed in the 1620s. Twenty years later its role was taken over by the 'cutt'. (This type may have originated by cutting the muzzle off a defective weapon, but very soon it became a type in its own right, and guns were ordered from the founders as 'cutts'.) In about 1670 Sir Anthony Deane developed another short fat weapon, which Samuel Pepys called 'punchinello'. An example of this survives in the

Royal Armouries. The 'punchinello' did not catch on, largely because admirals were gradually moving away from the mêlée tactics of the first part of the century, and fleets fought in increasingly rigid lines of battle, so that opportunities for close action were rare. Cutts survived in use until the end of the seventeenth century, but the Board of Ordnance lists suggest that old stocks were not replenished, and no comparable type was developed for many years. The carronade appeared in the late 1770s, when admirals were beginning to discard or modify the old fighting instructions which discouraged close action.

Probably the best known image of the carronade is of a 68 pounder on the forecastle of HMS *Victory*, as used at the Battle of Trafalgar. In fact this image is quite unrepresentative in several respects. In the first place the *Victory* had only two carronades, an unusually small number for a ship of the line. The typical 74 gun ship of that time had twelve 32 pounders, and six 18 pounders. The *Victory* herself was re-armed soon after the Battle of Trafalgar, and fitted with ten 32 pounder carronades. Secondly, the 68 pounder was actually quite rare in naval service. It first appeared in July 1780, and was tested at the range at Landguard. It attracted a great deal of interest, and early in 1782 it was used to arm the lower deck of the 50 gun ship *Rainbow*. In September, the *Rainbow* had an encounter with the French frigate *Hebe*, in which the French captain immediately surrendered when he saw the enormous size of the ball coming from the *Rainbow*. In the same year the Navy board recommended *allowing two 68 pounder carronades for the forecastles of all classes of ships capable of supporting them on the requisition of their captains, and 42 and 32 pounder for those of the lower rates.*[1] These were to allow ships *to annoy the rigging and sails of an enemy to a very great degree at the beginning of an action, and otherwise prove formidable against their masts and hulls at a greater distance.*[2] A list of July 1782 suggests that the 68 pounder was carried only by the *Rainbow.* A note at the end of the list mentions *ships ordered to be supplied with carronades, of which no account has been received of their being fitted.*[3] Of eight ships and sloops, three were to have 68 pounders. The *Romney*, another 50 gun ship, was to have two 68 pounders, the *Vigilant*, a 64, was also to have two, the *Egmont*, a 74 gun ship, was to have her whole lower deck fitted with twenty eight 68 pounders. It is not clear how many of them were ever fitted. William James mentions that two 74s, the *Canada* and the *Goliath*, were also fitted with a pair of 68 pounders each. The war ended before many more ships could have been fitted.

The idea of an all-carronade armament eventually became discredited after it was discovered that an enemy armed with long guns could stand off from the ship with carronades, and batter her at long range. The 68 pounder carronade also had its inherent disadvantages. At this time, the 42 pounder long gun was being phased out, because its ball was too heavy for a man in action to handle. This must also have applied to the 68 pounder carronade, and its rate of fire must have been very slow. No ships are known to have been fitted with the 68 pounder during the 1790s, and very few, apart from the *Victory*, in the decade that followed. A manuscript Navy list in the National Maritime Museum gives the armaments of all ships between 1807 and 1813, and the 68 pounder is carried only by two ships of the line, and several bomb vessels.[4]

The carronade as we know was not fully developed in one stroke. It is not always easy to follow the story of its development, because

unlike the Blomefield gun it was produced by private enterprise. Although the Carron Company papers are in the Scottish Record Office, they tell us virtually nothing about the way in which the carronade was designed. Elsewhere, there are plenty of drawings of carronades, and a few surviving guns, but none of the guns, and few of the drawings, are dated. However, it is clear that the early carronade, as used during the American War of Independence, was quite different in shape, size and mounting from the developed carronade of the Napoleonic Wars. The exact moment of transition from one type to the other is not clear, but it must have come at some time during the peace of 1783 to 1793.

The very earliest carronades are dated 1778, and were very small, only 18 inches long (Figure 1). They are heavily decorated for the standards of the period, and have trunnions instead of the loop which was used in the developed carronade.[5] The next precise date comes from a couple of drawings of 1782–3 in the Public Record Office.[6] They clearly show a short gun, still fitted with trunnions (Figure 2). Once this type has been recognised, it can be seen that it occurs in several other examples. There are four known survivors of this type. Two are in Dover Castle, though their provenance does not seem to give any clue to their age (Figure 3). Another was recovered from the wreck of the *Sirius*, lost in 1790. The fourth is in the Royal Artillery Museum at Woolwich, where it was catalogued as a mortar. This in itself is quite significant — the original design for the carronade seems to turn the mortar on its side to fire horizontally, while taking on some of the features of the naval swivel gun. A good, if undated, drawing of the trunnion carronade is to be found in the National Maritime Museum (Figure 4). A model of a landing craft or gunboat, mounted with a short trunnion carronade, can be seen in the Science Museum.[7] Although dates are rather scarce, it is not unreasonable to say that all the dated examples of the short fat gun with trunnions are pre-1790, while all examples of

Figure 1   Very early carronade in the Royal Armouries.

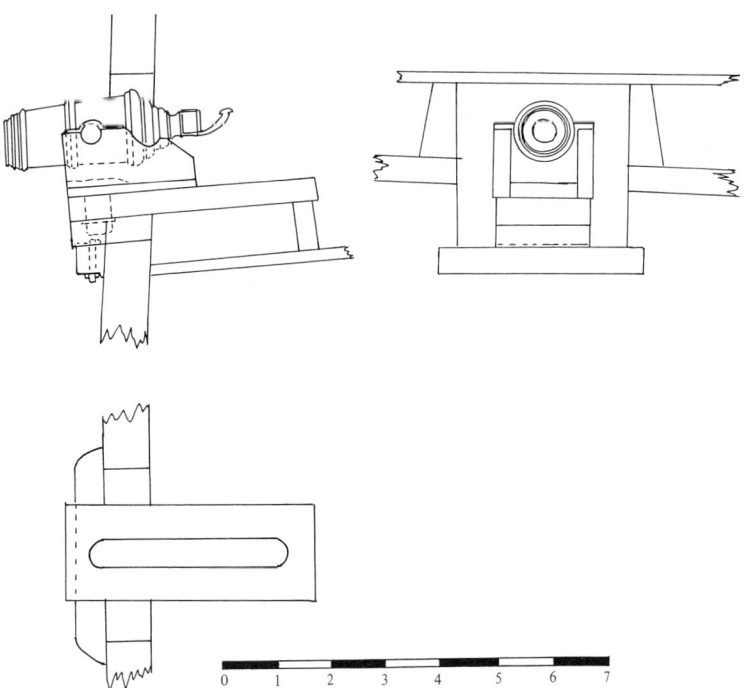

Figure 2   Carronade of 1782, based on drawings in the Public Record Office.

Figure 3   One of the carronades at Dover Castle.

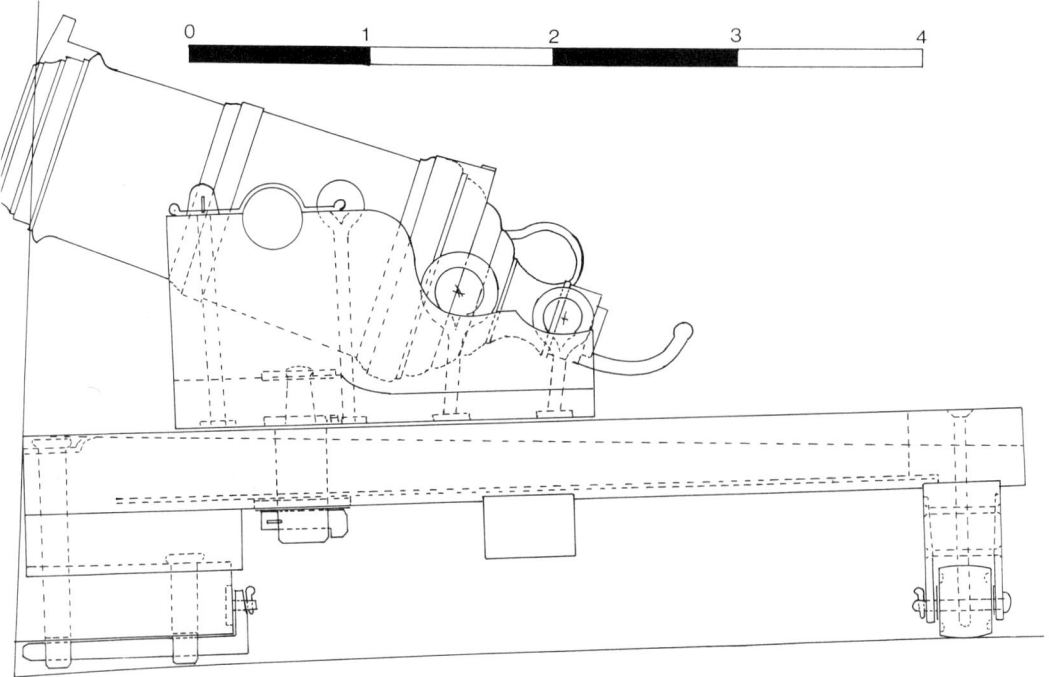

Figure 4   A short carronade, based on a drawing in the National Maritime Museum.

18

the later type of gun, with a loop and nozzle, occur after 1790.

It is worth noting that the number of carronades in use during the American War of Independence has often been underestimated. James' *Naval History* gives a table headed *A list of all the ships, down to 20 guns inclusive, so ordered to mount carronade as were in existence on the 1 January 1793.*[8] Only 40 ships are on the list out of a fleet of nearly 500, and the heading could perhaps be taken to mean that no other ships carried carronades. However, an earlier list from July 1782, in the Barham papers in the National Maritime Museum,[9] includes 167 ships and vessels, approximately one third of the fleet. If we allow for those ships which were too old for active service, and for those which had not been in a home dockyard for refit in recent years, it seems quite likely that about half the active fleet would have carried carronades by the end of the war. The ships on James' list are simply those which still survived ten years later. However, the list does tend to prove that no new ships were so fitted during the peace, and presumably production at the Carron works was very limited, if not totally stopped. It is probably during this period that the design was reconsidered.

Written records are not explicit about the development of the carronade, but they do give us a few clues. If we take the 18 pounder carronade as an example, we find that the gun was first tested by the Board of Ordnance in July 1779.[10] It was 33 inches long, and weighed 8 cwt. This trial did not impress the Board, but another took place a year later, with rather more success.[11] This time the 18 pounder was even shorter, only 29 inches in the bore. This is in line with the dimensions of the guns in the PRO drawings, and also with the undated National Maritime Museum drawing. By 1793, other types of 18 pounder had begun to appear.

The Ordnance Board had 100 of the 29 inch guns, along with five 24 inch guns (presumably experimental) in its stores. It also had 62 guns of 38 inches, and there are signs that these were beginning to supplant the shorter model.[12] In 1794, the Ordnance Board pointed out that it had no long 18 pounders left in stock, but 480 unappropriated short ones.[13] It seems that the older short pattern guns were being withdrawn from use, and longer ones issued. The Ordnance Board asked whether it was to issue the old guns, or scrap them. It is not clear what the decision was, but in 1796 there were only 211 short guns and 85 long guns left. The very short experimental guns had disappeared. In the same year, the officers at Woolwich were ordered to replace the old-pattern carronades of the sloops *Hornet* with new ones. They replied *there are no 12-pounder carronades of any other than the old pattern in the service, and consequently that order cannot be complied with.*[14]

The new pattern carronade had clearly made its appearance by 1793. The guns listed and drawn after that period usually conform in weight and dimensions to the gun that was in production for many years after the Napoleonic wars. Its barrel took on the characteristic shape and proportions, and it was mounted by means of a bolt and loop under the gun, rather than by trunnions on the side, as with other guns.

However, it is quite possible that at this stage certain other features had not yet been established. In particular, the hollow nozzle at the muzzle may not have been fitted to the earliest 'long' carronades. One surviving gun, mounted on the wharf at the Tower of London, quite clearly has the new shape, but has exactly the same muzzle as the old pattern gun, without the nozzle.[15] Taken on its own, it might be possible to dismiss this as an eccentricity, for the gun in question was made for a private owner, and was intended for use on land rather than at sea. However, there is another source that shows a very similar gun, but mounted for service aboard ship (Figure 5). This source is taken from the history of the Carron Company,[16] but it has not been possible to trace its provenance, as the author gives

Figure 5 A long carronade with a loop instead of trunnions, but without a projecting nozzle.

no sources. It appears to be part of a set with another picture, also undated, in the National Maritime Museum (Figure 6).

The second picture shows the Sadler pattern gun, developed as an experiment in 1796. This may give us an approximate date for the carronade drawing. The evidence is far from conclusive, but we must bear in mind the possibility that there was an intermediate type of carronade, without the nozzle, which was in use in the early to mid 1790s.

Furthermore, there is a possibility that for a time the nozzle was not hollowed out, but its inside simply continued as part of the bore. The evidence to support this comes solely from a few rather rough drawings, which are undated. It may, however, just be careless draughtsmanship. Furthermore, the length of the nozzle was never included in the total length of the carronade, suggesting that it was always hollowed out.

The purpose of the nozzle was two-fold. It would have helped to throw the flame of the

gun further out, increasing protection for the sides of the ship. It would also have helped to strengthen the muzzle of the gun, while enduring very little stress itself. The swell of the muzzle, found on more conventional guns, was therefore unnecessary.

If we consider the breech of the carronade, the picture becomes more complicated and the evidence even less conclusive. The basic profile on the breech is made up of three circular shapes. This can not be seen on the early gun of 1778, and it has only just begun to develop in the PRO guns of the 1780s. However, it can be seen clearly in all subsequent guns, including the drawing in the National Maritime Museum. Behind the breech itself, the short carronades seem to have had a cascable, which was almost cubical in shape.

Early guns also had a 'tiller' — a small handle extending behind the gun. This clearly shows the influence of the naval swivel gun. However, it is not easy to say what the carronade's tiller was used for. The swivel gun was very

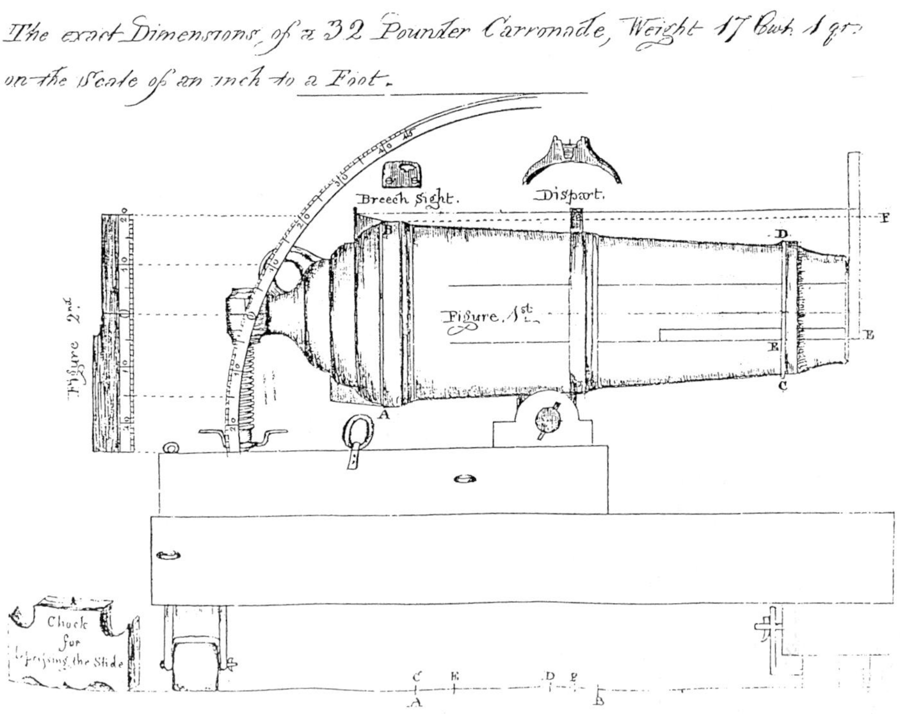
The exact Dimensions of a 32 Pounder Carronade, Weight 17 Cwt. 1 qrs
on the Scale of an inch to a Foot.
Breech Sight.
Dispart.
Figure 1st
Figure 2nd
Chock
for

Figure 6 A carronade with the nozzle, but apparently without the hollow inside, based on a drawing in the National Maritime Museum.

light — usually firing a half-pound shell — and was not allowed to recoil. A man could quite easily aim it with the tiller, and fire while still holding it. Clearly this was impossible with a carronade. Perhaps the tiller was used for lifting the breech, so that wedges could then be inserted under it, or perhaps it made traverse a little easier. Tillers are invariably shown in drawings, but it seems that they have not been found with the surviving examples. Presumably this means that the tiller was removable, and was seldom used in practice.

The tiller was soon supplemented by another piece of ironwork, mounted above the cascable. This was a ring, intended to help retain the breech tackle. Such rings can be seen on the *Sirius* gun, on the guns at Dover Castle, and also in the drawings in the National Maritime Museum. Unlike the tiller, the ring seems to have been cast as a part of the gun. It was much lighter than the breech ring later developed for the Blomefield gun. The ring tended to get a little heavier over the years, though it was evidently not fitted to all the guns of the 1800s.

Another type of handle appears on the Tower gun and in the gun in the Carron Company book. This curves round the back of the breech, and is attached to it on both sides. The store lists of the 1790s often include 'carronades with handles', but the handle fell into disuse soon afterwards.

The elevating screw, often regarded as a characteristic of the carronade, made its appearance at about the same time as the gun was redesigned. In fact, it was not a totally new idea, for it had been suggested by William Bourne in 1578.[17] The elevating screw naturally involved a re-design of the cascable; usually it was round in plan, and square in profile. Around 1790, guns were also moulded with a small wedge above and below the breech. The upper wedge served to hold some type of gun sight, while the lower allowed the use of wedges instead of the elevating screw. The screw must have been quite slow to operate, and wedges were presumably provided as an alternative.

The carronade was probably not fully developed until about 1795 — with a loop instead of trunnions, a nozzle, elevating screw, and all the features we have come to associate with it (Figure 7). It was just in time for a period when close action was becoming more fashionable than ever before.

The development of the Blomefield gun has to be approached rather differently. The gun was developed under the authority of a government department, largely by Thomas Blomefield, who was a prolific letter writer. Much of his correspondence has survived, either in the Public Record Office or in the Tower of London. H A Baker used Blomfield's correspondence to trace the story of his reorganisation and improvements made in the artillery department. This paper will look at some aspects of his contribution to gun design.

Thomas Blomefield was 36 years old and a captain of artillery, when he was appointed to the new and relatively junior post of Inspector of Artillery in 1780. He remained in the post for the rest of his life, during which he became a baronet and a full general. Although an army officer, he had considerable experience at sea. In his youth, he had spent a short time in the navy, but had entered the Royal Military Academy by the time he was 14. He served as artillery officer in bomb vessels, and took part in many amphibious operations. He was clearly familiar with the problems of naval gunnery.[18] Though he is well known for his insistence on high standards in the casting of guns, his knowledge of naval conditions made him aware that absolute perfection was not realistic aboard ship. *Was not some latitude allowed, the number [of guns] rejected would render it impossible for the contractors to furnish them on the same terms ... The nature of [the naval service] renders it unnecessary to insist upon the most scrupulous exertions in every part*

of their contract.[19] Since a naval gun was fired from a moving platform, it was inherently inaccurate, and perfection in gun-founding would be largely wasted.

Blomefield, of course, was not the first person to design a standard pattern of gun in Britain. In the seventeenth century, the proportions of a gun had been left largely to the skill of the craftsman who made it, in the same way as a ship was largely designed by the master shipwright, even in private yards. In both cases, the early eighteenth century demanded greater uniformity and higher standards. In guns this was imposed by Borgard. After Borgard, the Armstrong pattern gun survived until Blomefield's time.

Borgard and Armstrong had tended to codify the practices of the time, rather than introduce any new concepts. Blomefield, on the other hand, was concerned with improvement. One factor that caused him initially to look at gun design was the invention of cylinder powder.

In 1783, it was suggested that charcoal, which was charred in cylinders or ovens rather than in kilns, would burn much more uniformly. Powder made from this type of charcoal was considered much more powerful than the conventional type. With the old type of powder, the charge was from 40–66% of the weight of the ball, according to the size of the gun. When cylinder powder was eventually introduced to the navy around 1800, 33% became standard. The first proofs with cylinder powder were carried out in 1791, although Blomefield had already done much of the design work on his new guns by that time. The second factor was the failure of numbers of guns under proof. This aspect has been covered by Baker.[20]

Blomefield began his reform of gun design in 1786. He was largely given a free hand by his superiors on the Board of Ordnance, and allowed to produce guns to his own specification. He developed a special relationship with one iron company, Samuel Walker of Rotherham, and Walker was entrusted to produce the experimental guns. The first new

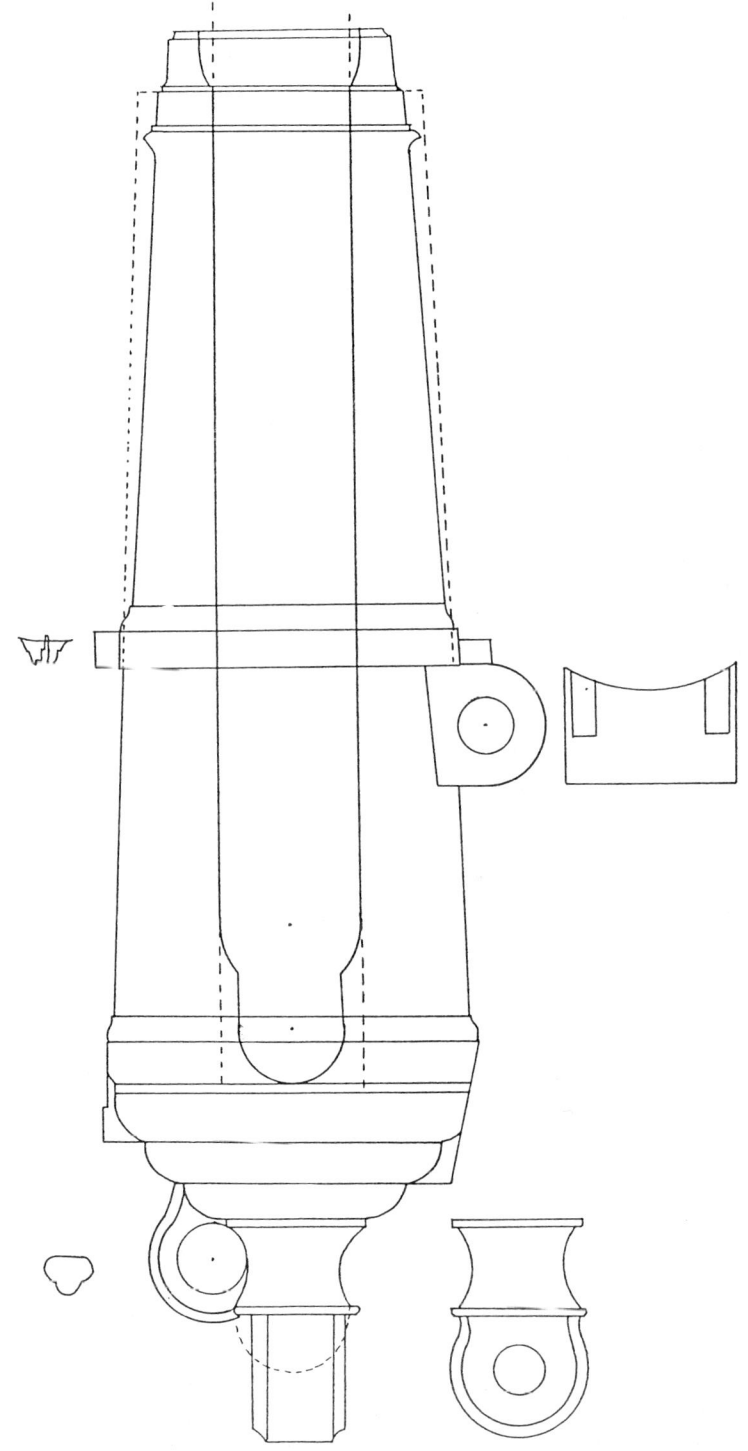

Figure 7   The fully developed carronade, based on drawings in the Carron papers in the Scottish Record Office.

design proposals came in May 1786, when Walker was asked to cast two 18 pounders to a new pattern. They were landed at Woolwich in late June, and subjected to proofs. The proofs were reasonably successful, but Blomefield wanted higher standards. In July he *reported his opinion that the guns might be rendered still lighter and may also pursue a greater degree of resistance than those of the old construction, that he had prepared a draught of a gun upon that principle.*[21] The draught of the gun was to be sent to Walker, who was to cast two 32 pounders and several 18 pounders to this pattern. Walker retained his commercial awareness. He wrote to Blomefield, *The price at which we have taken the contract will not admit of our doing much out of the common road, the limits to which we are convinced are very narrow, and casting and boring is not attended with much more risk than formerly . . . . We doubt not you will recommend it to the honourable Board to make us some allowance further than the price at which we have taken the contract.*[22]

Blomefield was to alter gun design in three main ways: he made the breech more rounded, so that the metal was more evenly distributed to resist the explosion of the charge; he redistributed the weight of metal along the length of the gun, to the advantage of the part nearest the breech, where the charge was exploded; he added loops to the breech, to hold the breeching ropes.

The breeching ring, or loop, was probably intended to make it easier to fire the gun at an angle to the side of a ship. In the older guns, the breech rope had been spliced over the cascable of the gun, so that the length of rope on each side was fixed. If the gun was fired at an angle to the ship, one side of the rope would have to take the whole force of the recoil, putting it under great strain. On Blomefield guns the rope was not fixed to the gun, but merely passed through the loop, so that it was free to run. Presumably both ends of the rope would act together in restraining the recoil. Blomefield, himself, placed great

importance on allowing the rope a free run through the loop.

In December 1786, Blomefield wrote to Walker asking him to avoid any projections on the inside of the loop which *might gall or impede the free passage of the breechings when the recoil of the piece is checked by them.*[23] Clearly, the breech rings were more useful at sea than on land, where it was not always necessary to restrain recoil in this way. In January 1787 in the early stages of development, the guns with loops were referred to as 'sea service guns' — *the Board of Ordnance wish to have loops on all sea service guns, and have wrote to the founders to cast them accordingly.*[24] However, the Blomefield gun was soon adopted for fortress use, and in practice there was no distinction between land and sea ordnance.

The casting of rings caused some difficulties to the gunfounders in the early stages. Bronze guns had long been fitted with loops above the chase, mainly for lifting the weapon in and out of a ship or fortress. However, iron cools much more quickly than bronze, so there was a danger of it setting before it had filled all the spaces in the mould. Walker had serious doubts about the practicality of casting loops in iron, and told Blomefield so. *It will most certainly be a considerable inconvenience and risk to make guns with loops — more so in the boring than merely in the casting.* Blomefield replied that he should persevere, and eventually it proved successful. Within a few months, Walker had found new methods. *We have hit upon a mode of casting and boring guns with loops on the cascables, which will enable us to do them more readily than we at first imagined — there will, however, be more risk and trouble than without them.*[25] The casting of rings was made somewhat easier by the new custom of boring out guns from the solid. In the past, the bore had been cast with the gun, but casting in the solid allowed the metal to reach the breech and the ring more freely.

The redistribution of metal around the breech of the gun was clearly intended to

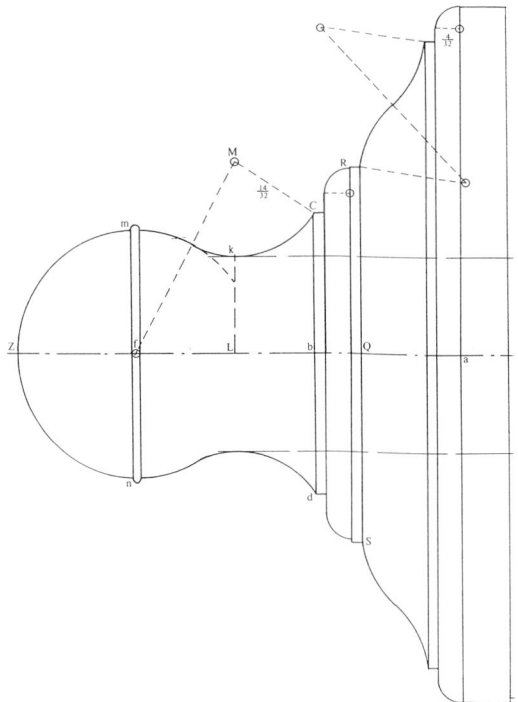

Figure 8   The method of drawing out the breech of an Armstrong pattern gun, from a manuscript in the National Maritime Museum.

Figure 9   The breech of a Blomefield pattern gun, in the Royal Armouries.

eliminate the weak spots of the old pattern. With an eye to decoration rather than efficiency, the gunmakers had allowed the base of the piece to be adorned with rings and reverse curves, leaving several obvious sources of weakness. Blomefield drew the base of the gun as a single curve, whose centre was located in the chanber where the powder exploded (Figure 8). The whole base now presented a uniform thickness to resist the blast. As described in his letter of December 1786, *In moulding the breech of the 32 pounders, the convexity should fall in with the sweep of the hinder part of the base ring as smoothly as possible, in the same manner as in the 18 pounders* (Figure 9).[26]

The third alteration Blomefield made was to remove metal from the chase and muzzle of the gun, and add it to the first and second reinforces where the charge was exploded. He wrote in one of his very first letters on the subject, *As these pieces are strengthened in these weak parts near the charging cylinder where the straining of the charge chiefly lays, I can have no doubt of their resistance, at least to any proof which those of the present construction are capable of withstanding.*[27] Since a gun was cast with its muzzle uppermost, the relative thinness of the neck of the piece caused difficulties for Samuel Walker. When shown the draughts for Blomefield's first experiments, he commented, *We rather fear that these guns, from their shape, will be more liable to holes than the old ones, owing, we conceive, to the diameter behind being rather increased and decreased towards the muzzle.*[28] He wrote again in July, *The general shape of the gun is very proper for the sinking of the metal. The disadvantage is from the smallness of the neck, which cooling first and fixing prevents the lower parts of the gun being fed with liquid metal from the head, and cause the parts near the neck to be porous.*[29]

The earliest known set of dimensions for a Blomefield gun are from August 1786.[30] The main changes compared with the old pattern Armstrong gun was indeed to reduce the diameter of the neck of the piece, from 13.8 inches to 13.2 inches. The diameter immediately in front of the base ring was in fact slightly reduced, while that just behind the second reinforce ring remained almost constant. The Blomefield gun was slightly

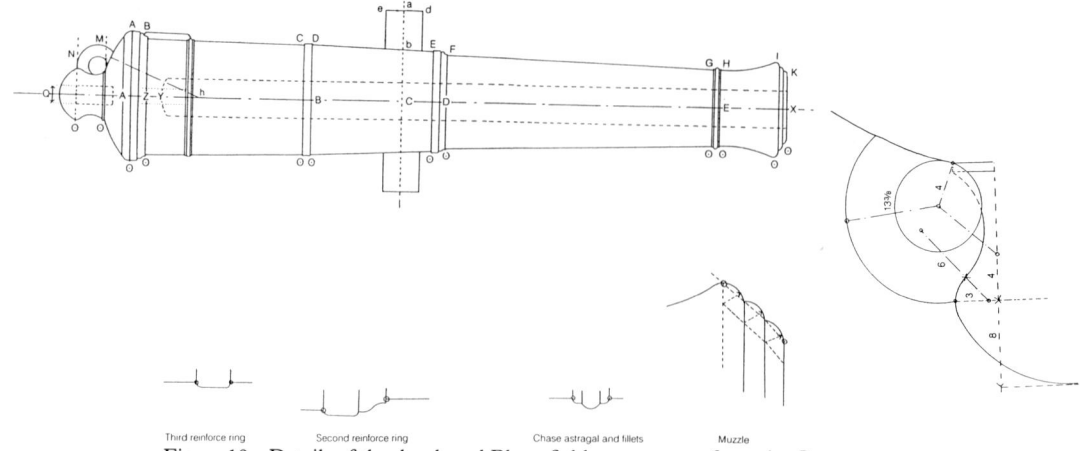

Third reinforce ring   Second reinforce ring   Chase astragal and fillets   Muzzle

Figure 10   Details of the developed Blomefield pattern gun, from the Carron papers.

wider immediately in front of the second reinforce ring, so that the narrowing of the gun at that point was somewhat reduced. The swell of the muzzle was also considerably reduced, from 17.7 inches to 16.97. There was a reduction in diameter along most of the length of the gun, so that it must have been considerably lighter. This indeed was one of Blomefield's early aims. In July 1786 he expressed the opinion that *the guns might be rendered still lighter and may also preserve a greater degree of resistance than those of the old construction.*[31]

However, this was not to be. Walker and Blomefield continued their experiments during the next few years, but Walkers guns seem to have failed at proof as much as anyone else's. Most of the weight that had been taken off the Armstrong pattern was added on again during the experiments. By the 1790s, after the Blomefield gun had completed its development, the forward part of the gun was almost the same as it had been in 1786, except that, in 32 pounders, another half inch had been taken off the diameter at the swell of the muzzle. The diameter forward of the breech ring had gone up to 20.12 inches, slightly more than that of the old Armstrong pattern. That of the second reinforce had gone up from 17.21 to 17.38 inches. This meant that the narrowing at that point had increased from approxi-

mately 0.9 of an inch to 1.08 inches. It was therefore slightly greater than it had been in the old pattern guns. Unfortunately, this period of the gun's development seems to be less well documented than others, so it is not possible to be certain about the reason for these changes. However, it seems likely that they were introduced by the need to make the gun stronger for cylinder powder.

Blomefield also redesigned the muzzles of guns. Instead of the ogees and reverse curves of the old pattern, he used three curves to reduce the muzzle from the swell to the face. This is not likely to have improved the efficiency of the gun, but it does tend to give the Blomefield guns a more 'business-like' appearance than their predecessors.

Another improvement was in the system of aiming the gun. As early as January 1787, Blomefield was ordering that *sea service with loops are to have square pieces of metal on the chase.*[32] Presumably this was intended to help with sighting. Later, degrees were marked on the breech ring, with small notches cut in the swell of the muzzle to line up with them (Figure 10).

It is not quite clear exactly when the Blomefield gun stopped being experimental, and became the standard weapon. By the Spring of 1787 the basic shape had been evolved, but the exact proportions were still not settled. The letter books are not explicit, and the issue is confused by the fact that Walker was also experimenting with different types of iron at the same time. The initial experiments were carried out with 18 and 32 pounder guns, and different types were added to the repertoire. Other gunfounders were gradually brought into the scheme, and were sent draughts of the new pattern for future casting. By 1792 there was reference to 'new iron ordnance' which it was proposed to cast.[33] In 1794, old pattern guns were still being received at Woolwich, and two 32 pounders burst on proof by *separation of the breech from their bodies*.[34] They were now considered too weak to withstand cylinder powder, and no more were to be received. However, one exception was made two years later. Messrs Gordon and Stanley received *permission to deliver 24 pounders of the old pattern, which nature of ordnance at the time the offer was made were much wanted, but are now no longer so; I beg leave to recommend that they may receive directions to deliver none in future, but those of the new pattern.*[35]

The gun lists of 1808[36] suggest that a few old pattern guns were still in use at sea, but were being withdrawn as quickly as possible. The lists also show that most ships had Blomefield pattern guns, and carronades. By that time, the Royal Navy had nearly 1000 ships, needing about 30,000 guns and carronades. Most of these had been cast since 1790. The carronade and the Blomefield gun, as well as winning the Napoleonic Wars, therefore made an enormous contribution to the development of the British iron industry.

Notes

1 H L Blackmore, *The Armouries of the Tower of London, part I, Ordnance*, London, 1976.
2 Public Record Office Adm series, Admiralty to Navy Board, 21/1/82.
3 National Maritime Museum MID/9/2.
4 National Maritime Museum RUSI/64.
5 op cit, note 1, pp. 144–5.
6 Public Record Office Adm 106/3472.
7 Science Museum, inv. no. 1865–41.
8 W James, *Naval History of Great Britain*, vol. I, p. 423.
9 National Maritime Museum MID/9/2.
10 Public Record Office WO 34/147.
11 Public Record Office Adm 7/940.
12 Public Record Office Supply 5/54.
13 Public Record Office Supply 5/53.
14 Blomefield letter books in the Royal Armouries, no. 5, f. 100.
15 op cit, note 1, p. 145.
16 R Campbell, *Carron Company*, Edinburgh, 1961.
17 W Bourne, *The Arte of Shooting the Great Ordnance*, 1587.
18 Dictionary of National Biography, vol. V, p. 208.
19 Public Record Office Supply 5/52, 28/8/96.
20 H A Baker, *The Crisis in Naval Ordnance*, published by the National Maritime Museum.
21 Public Record Office WO/47/108, 25/7/86.
22 Public Record Office Supply 5/49. 24/8/86.
23 Public Record Office WO 47/108, 29/12/86.
24 Blomefield letter books no. 3, 10/1/87.
25 Public Record Office Supply 5/49. 24/8/86, 11/10/86.
26 Public Record Office WO 47/108, 29/12/86.
27 Public Record Office WO 47/108, 13/7/86.
28 Public Record Office Supply 5/49, 31/5/86.
29 Public Record Office WO 47/108, 17/7/86.
30 Public Record Office Supply 5/107, 21.8.86.
31 Public Record Office WO 47/108, 25/7/86.
32 Public Record Office WO 47/108, 18/1/87.
33 Blomefield letter books no. 5, 31/8/92.
34 Public Record Office Supply 5/43, 11/1794.
35 Blomefield letter books no. 5, f. 109, 28/8/96.
36 Public Record Office Adm 160/154.

The Introduction of the Carronade into Dutch Naval Service in the Late 18th Century

J P PUYPE

After years of mismanagement, the last quarter of the eighteenth century saw a virtual renaissance of the Dutch fleet. Within a period of twelve or thirteen years, from 1777 till 1789, no less than forty-six ships of the line armed with 50 to 76 guns were built. Nine ships had seventy-four guns, twenty-nine had 60–68 guns and eight had 50–56 guns.[1] In addition, ten frigates of 36 guns were built as well as twenty-six lesser men-of-war, not counting a number of cutters and other craft (Figures 1 and 2).

There were 92 new warships, most of which were laid down between 1779 and 1784. During this period there was a war between the Dutch Republic and Britain (1781–1784), known in Holland as the 'Fourth Anglo-Dutch War'.

The expenditure for this building programme was 75 million guilders. This was an astronomical sum to which, as usual, the province of Holland contributed the largest part.

According to the 19th century naval historian De Jonge, a Dutch contemporary to James, the author of *The Naval History of Great Britain*, the zest of the era was mainly fed by the desire to *lick the English opponent*, and teach him the lesson that Dutch sailors were as brave as their forebears, more than a century earlier.[2]

It is true that after years of neglect ship-building advanced at too fast a pace. During the war of 1781–1784, the Baltic, the traditional Dutch source of timber, was practically cut off and the Dutch yards had to use young, fresh wood supplied from middle Europe via the Rhine. Moreover, the naval arsenals, after decades of neglect, were depleted of many stores which could not be reinstated at short notice.

However, the Dutch fleet was expanded. The admiralties of the Dutch republic, of which there were no less than five, discarded the traditional, clumsy type of warship of the early 18th century. Instead, they made good use of the big strides in shipbuilding which had taken place in France and Britain. Not only were the Dutch ships bigger and stouter than before, but they featured better lines as well.

The Amsterdam ships, especially, excelled. This was largely due to the restless industry of the able chief of the Amsterdam naval shipyard, an Englishman — of all people — by the name of William May. He was the son of one of the three English shipwrights, who had been commissioned in 1727 by the Amsterdam admiralty to improve the way in which ships were built.

Another improvement introduced at this time was the coppering of ships below the waterline.[3]

Armament, gunnery and tactics were also improved, mainly through the influence of a few highly-motivated professional officers in essential posts. The same names occur regularly in the many contemporary texts on artillery: Van Kinsbergen, Zoutman, Hartsinck, May, Van Byland, Van der Hoop and a number of others.

Van Kinsbergen, however, was the driving force behind the improvement in naval artillery.[4] He wrote several classic works on various naval subjects including gunnery and tactics, witnessed trials with gun or carriage improvements or with new guns, and also introduced a number of practical features for cannon and carriages (Figure 3).

Figure 1   The man-of-war *Landscroon*, 44 guns, watercolour by J P van Braam being the frontispiece of a manuscript from 1762 containing inventories and other lists of this ship. The *Landscroon* is seen here carrying every sail she could carry, including an extra mast with sails instead of the ensign flag staff. She was built in 1755 in Amsterdam by Charles Bentham, yet still seems to feature pretty much 'Dutch' lines. (Stolen from the Nederlands Scheepvaart Museum in January 1988).

Figure 2   The two-decker *Staten Generaal*, depicted twice. Oil painting by E. Hoogerheijden, 1795. This vessel, of 76 guns, was built in Rotterdam by P Glavimans and it featured the better lines which the newly-built Dutch warships had by the last quarter of the 18th century. The *Staten Generaal* measured 180 feet and was commissioned in 1793.

Figure 3   J H van Kinsbergen, 1735–1819, depicted in the uniform of a captain in the Dutch navy in this engraving by R Vinkeles of 1781. An innovator in many fields, especially artillery and a veteran of many battles, including service in the Russian navy. He became commander in chief of the navies of Holland and Zealand in 1793, and of the entire Dutch fleet in 1795.

For example, Van Kinsbergen invented what we call the 'traversing chock'.[5] This was a rounded piece of wood which, before action, was inserted between the cheeks of the carriage at the front to enable it to be easily traversed while resting against the ship's side (Figures 4 and 5). This prevented the problem of a wheel crashing against the side while running the gun out when the ship heeled over or lurched. More often than not, the wheel, or rather the axle, would break and the gun be rendered useless. The introduction of the 'traversing chock' overcame this problem. It was taken up by other countries, initially by the French navy, who first officially introduced it on carriages of the 1786 pattern, and later by the Royal Navy, who, however, did it the other way round and gave the bottom sill of the gun

ports a rounded protrusion on the inside of the ship's wall.

In 1790 Van Kinsbergen, as rear admiral of a Dutch naval squadron, visited Spithead. The Dutch commander enthusiastically reported to the Stadtholder, Prince Willem V, that the renowned British admiral Lord Howe had praised the Dutch ships for their order and precision. Lord Howe had observed that he would have deemed it an honour to have under his command such able and experienced officers as those in the Dutch squadron.[6] This was, of course, an elegant and non-committal observation, yet, nevertheless, a recognition of the obvious improvement in the Dutch navy. Armament was improved by the introduction of newer patterns of cannon, as well as by arming the ships more heavily. It was for this reason that the carronade was introduced into the Dutch navy.

The carronade was a type of hybrid gun — half a cannon, half a howitzer, short and light for its calibre. It had been designed by General Robert Melville and was produced by the Carron factory in Falkirk, Scotland.[7] The carronade was designed as a means of inflicting serious damage, but at the same time to be inexpensive to produce and lightweight. The disadvantage, however, was its feeble range. It seems to have been originally developed for arming merchant vessels, but was soon taken up by the Royal Navy. The Board of Ordnance, however, does not seem to have given it its approval.

The earliest carronades (not counting various prototypes) were developed in 1779. According to James, there were already well over 400 English men-of-war equipped with carronades, as an additional armament by 1781 (Figure 6).[8]

The Dutch must have been aware of the new type of gun at an early stage, as, in 1780, Captain Baron van Knickel of the Admiralty of Zealand brought two examples of carronades with him to the Netherlands.[9] I have not yet been able to ascertain whether Van Knickel acted under orders or whether he acted on his

Figure 4   Bronze eight-pounder of the Admiralty of Rotterdam, 1783, lying on a reconstruction 12-pounder sea carriage of the pattern of 1800 of the navy of the Batavian Republic. Note the traversing chock between the
cheeks at the front invented by Van Kinsbergen.
(Nederlands Scheepvaart Museum, Amsterdam).

Figure 5   Close-up of the traversing chock.

own initiative. It is, however, definite that contact had been made between the Carron foundry and one of the Dutch admiralties. This is proved by an interesting reference, dated 27 August 1779, in the Scottish Record Office, Edinburgh, where the company's records are kept.[10]

The reference concerns the supply of guns and shot . . . *for the use of one of your ships now fitting out at Harlingen.*[11] By order of a John May, most probably the father of the afore-mentioned William, the Carron Company shipped the following aboard the *King of Spain*: twenty-six 24-pounders, twenty-six 18-pounders and six 6-pounders, together with almost 5,000 pieces of shot.

On 13 June 1780, probably at the instigation of Van Kinsbergen, the Stadtholder issued an order that the two carronades brought over from Britain should be put to test. This test was carried out during the summer of 1780 and a report, signed by Hartsinck, Van

Kinsbergen, Zoutman, Pichot and May, was submitted on 8 September.[12]

The two carronades, an 18-pounder and a 24-pounder, were tested against two iron cannon with identical bores. The long guns were said to have been ... *cast in Sweden*, and were probably of the so-called Finbanker pattern or, rather, one of the several Finbanker patterns, as they were the standard-issue iron guns in the Dutch fleet.

The report does not specify how the guns and the carronades were mounted. They were fired at two sections of a ship's side placed on the beach at a distance of 305 paces (about 230 metres). The targets, each measuring 21 feet wide by 19 inches thick, were placed at a distance of 45 feet from each other.

The guns were loaded with standard shot and with heavy-powder charges of one-third and one-quarter of the ball's weight. It appears that all shot penetrated the targets. A test with a lesser-powder charge in the carronades, set at an elevation of $1-1\frac{1}{2}^{\circ}$, revealed that the bouncing shot still penetrated the ship's sides. The report concluded, rather ambiguously, that ... *all guns had properties that could be used to good advantage by the land.*

The war between England and the Netherlands prevented the delivery of any carronades to the Dutch republic. Nevertheless, contacts between the Admiralty of Amsterdam (through John May) and the Carron Company continued on a modest scale.

There is an interesting correspondence between the high commissioner of the Admiralty of Amsterdam, Van der Hoop, and the director of the Carron Company, Charles Gascoigne. This is now kept in the General Archives of the Netherlands in The Hague.[13] One important document among these papers dated 29 November 1783, concerns Gascoigne's authorisation of three Amsterdam merchants to act in the name of the company and obtain delivery contracts from all the Dutch admiralties, though first and foremost that of Amsterdam.[14]

On 17 May 1782, Gascoigne had written to one of the three merchants, a certain Ten Cate, supplying him with sketches and a short description of a carronade:

Carronades are made to 7 calibres of the [eir] length, the chamber included, weighing about 60 lb iron for each pound of the weight of the shot. The bore is chambered for the powder charge, having the size of the next lesser caliber, meaning [for instance] that a 68-pounder has a 42-pounder chamber, and of such length that it may contain an ordinary powder charge loosely bound in a flanel cartridge equalling one-twelfth of the weight of the shot for the lesser calibers and 1/14th for the bigger. The shot must be placed directly against the charge without the use of a wad since that gives more power to the ball and there is less danger of a fire anyway. Once loaded, the carronade is wadded in the usual manner. I must warn against the use of too heavy charges. If the carronade, because of prolonged firing, becomes hot (which is not likely because of its shortness), it will be necessary to slightly elevate the rearmost part of the lower carriage, but you should observe if the loaded piece can still move within the port. The windage and vents of the carronades are considerably less than those in other guns. Therefore you cannot use priming powder, but have to apply a quick match instead which has been found to be reliable on every occasion. Using both the flanel cartridges and the match, you prevent spilling powder on the deck as well as the burning out of the vent when the gun is fired. I propose to introduce hand locks as they are in use at Carron. These hand locks are attached to the rear of the piece and they feature an aperture in their bottom which is placed on the quick match and the gun can be fired while aiming it.[15]

The most interesting feature mentioned in this letter is the use of quick match, a fact not mentioned by later writers to my knowledge. The reference to the small windage is also of interest because the much narrower clearance

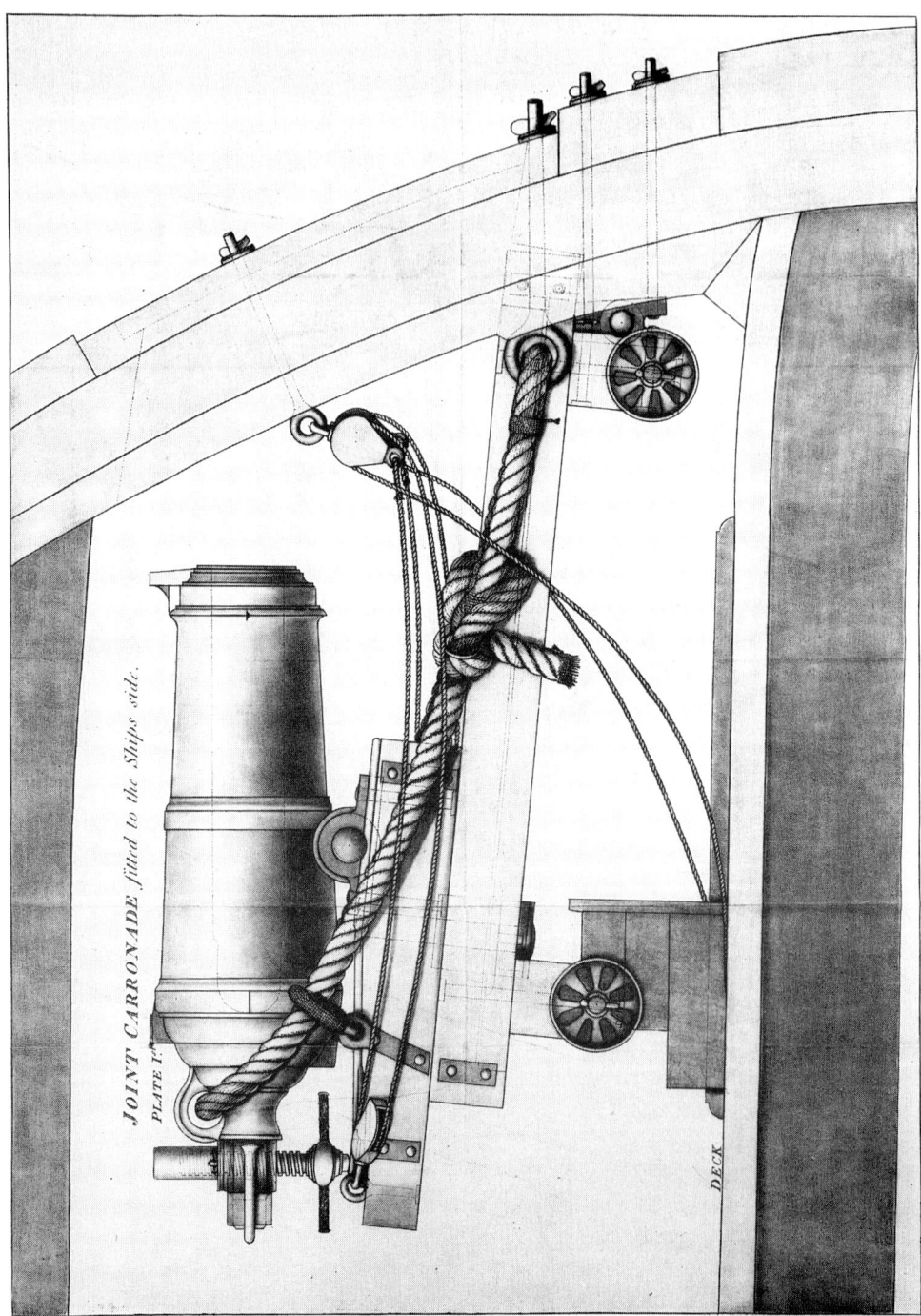

JOINT CARRONADE fitted to the Ships side.

PLATE 1st.

DECK

Figure 6 'Joint Carronade fitted to the Ships side, Plate 1st.'. This print is the earliest illustration of any carronade found in eighteenth-century Dutch records. This particular copy is among the file on carronades sent in May 1785 to the Stadtholder by the Dutch envoy in London. (Collection H.M. the Queen of the Netherlands, Royal Household Archives, Inv. No. A18–291 III).

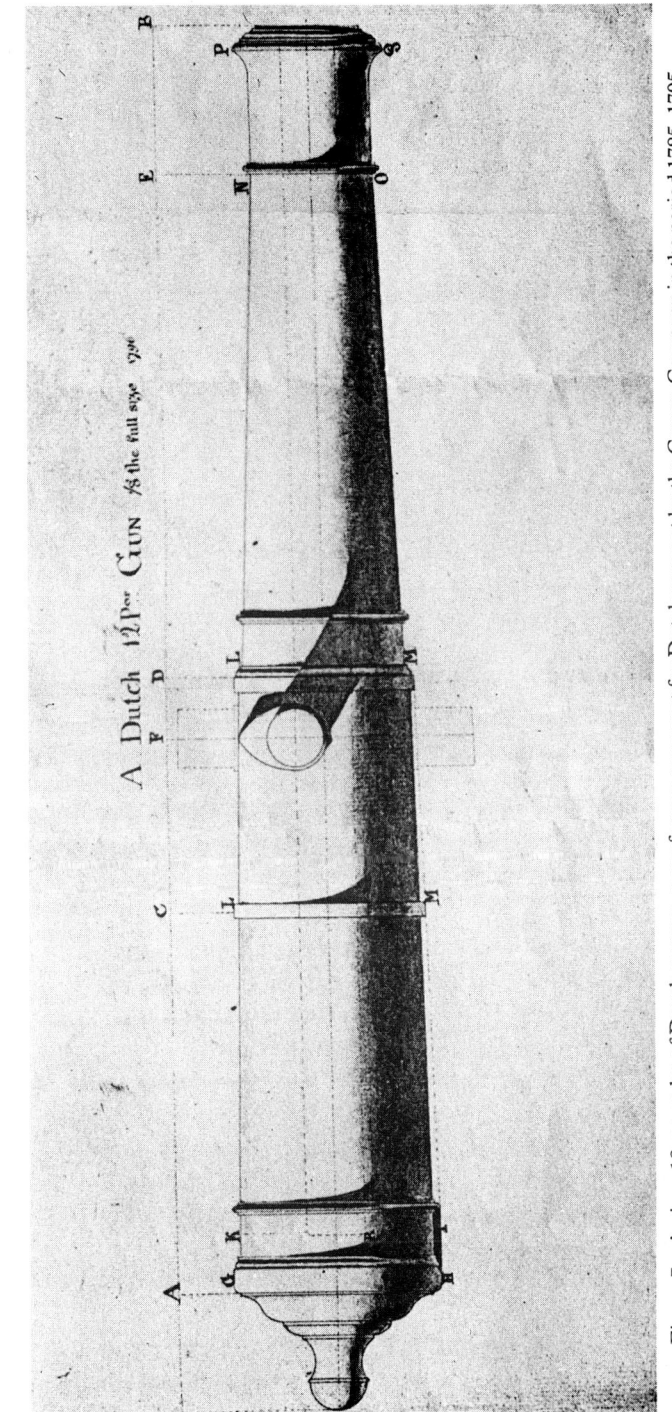

A Dutch 12 Por Gun ⅛ the full size 1790

Figure 7   An iron 18-pounder of Dutch pattern, one of many guns cast for Dutch account by the Carron Company in the period 1785–1795. (Scottish Record Office).

of the carronade paradoxically resulted in its ballistic results being far superior — at short range, of course — than those obtained with the standard long cannon, from which too many powder gases could escape past the shot during the initial split second after the explosion of the charge.

On carronades, the inside of the bore at the muzzle was hollowed out to provide space for the fingers when the shot was inserted. This is a practical piece of evidence for the much reduced windage; the clearance on standard long guns was so large that a cannonball could be eased into the muzzle with the fingers under it.

In his letter to Ten Cate, Gascoigne also gave several suggestions and rules which he thought should be observed in order to use the carronade to its full advantage. For instance, the gunner should bear in mind that the breeching rope, which was subject to stretch, should always stop the sliding carriage before the vertical connecting bolt could touch the end of the travelling slot in the low carriage. If the breechings were stretched too much, the bolt would smash against the end of the slot with such momentum that it could easily break.

Gascoigne also advised that the rear end of the lower carriage should always be elevated by 3° by means of a quoin, though wisely, he does not elaborate on this difficult procedure during action.

He also advised that when the ship was heeling over heavily, the leeward carronades should have their rear wheels removed, allowing them to rest on their axle blocks. Double-shotting, which was common with long guns at close quarters, was prohibited because it would inevitably result in too heavy a recoil and the carriages would break after a few shots.

That this was no exaggeration is supported by a memorial of January 1784 by the Count van Byland, who states that . . . *the Spanish navy uses since a certain time many light and short pieces from the Carron Company, the carriages of which broke by the violence of the recoil.*[16]

However, when reading this we should not forget that, at the time, there was much opposition to the use of carronades, both in the British and in the Dutch fleets. Count van Byland's remark, therefore, could also be interpreted as an attempt to rouse popular feelings against the new gun.

In 1782, for instance, it was reported that the . . . *Captains Duncan & Faukoner [sic] have always been great enemies of the carronades . . . they said there was no doubt of their standing a single shot, but that, upon their being heted, they would jump and kick about so confoundedly, that there would be no managing them . . . .*[17]

A Dutch naval officer, in a note from the same year, states that, . . . *the carrying of carronades by the ships, not as pieces in the batteries, but as an additional firepower, is most useful, especially light carronades of 12 lb in the fighting tops which could fire grapeshot etc. at the enemy decks . . . If the manufacture of twelve-pounder carronades for the Dutch fleet would be approved, it would be necessary to order, too, the sliding carriages and other parts needed in order to operate the guns from the ships [decks].*[18]

Carriages are mentioned in another letter by Gascoigne, dated 9 February 1784. It is here that reference is made to the first official order of carronades for the Admiralty of Amsterdam.[19]

It appears that initially 18 carronades of 24 lb were ordered. Apparently, the customer had requested a quotation per piece, with and without its carriage. The cost of the complete gun was £24, whereas for the piece without its carriage it was £18.[20]

Gascoigne noted, however, that it was difficult to supply the parts on their own, especially the iron carriage mounts which were also made by the factory. He then added that the prices were very competitive anyway, because they included proving the guns as well. Incidentally, as there was no government-appointed proofmaster in Carron, the guns would have to be sent to Woolwich Arsenal where the Board of Ordnance proofed guns. Instead, Gascoigne suggested the carronades for Holland be proved by the company's own proofmaster, under the supervision of an artillery officer from Edinburgh Castle.

Gascoigne's final comment as to why the company should deliver the carriages as well, was that it would be imprudent for the Dutch to make the mounts themselves, since he would not then be in a position to guarantee the quality of the iron.

Gascoigne then said that the eighteen carronades, complete with their carriages, would be ready within the time stipulated in the contract, adding that he had received precise and full information on the dimensions of the gun ports. He claimed he would do his utmost to deliver the carriages according to the exact specification . . . *in order to encourage further commissions.*[21]

Another passage in the letter indicates that Gascoigne would have preferred a larger order: *We shall have the 18 carronades tested in presence of Major Andrew Fraser, First Engineer of Scotland, who is well-known having been present at the former peace of Dunkirk to watch over the fortifications. The proving of so little a number of pieces will certainly involve great expense, but will be carried forward nevertheless.*

The records in The Hague also contain a concise report, by a John Zuil, to Gascoigne, dated Liverpool 23 January 1779. They concern the test of 18 lb carronades on two oak ships' sides as targets. He comments that the firing had been very quick, at ten shots within two minutes.[22]

The war between England and the Netherlands made it very difficult for Gascoigne to encourage further orders: *Your proposition to send an assortment of carronades, is not within our power to execute. Our government forbids us to do any consignation.* On the other hand, Gascoigne did all he could to ensure that his customers got a good impression of his products. Along with the letter he sent reports of trials with heavy carronades and he mentions, as a reference, Sir Charles Middleton, Controller of the Royal Navy and president of the navy Board. He adds: *We shall also send you for reading our letters from Spain and other correspondence on the commissioning of carronades. We are now busy finishing an order for 180 tons of carronades, mortars and bombs for Constantinople.* The true merchant in him is revealed in his last sentence: *And if Your Honour would prefer normal guns, we can always provide you with light pieces against very low prices* (Figure 7).[23]

In an earlier letter to Ten Cate, Gascoigne recommends that the heavily-laden cargo vessels of the Dutch East India Company would benefit greatly by carronades as part of their armament. This would result in a weight saving of three quarters of the weight of the normal ordnance and still give the same firepower. He further communicated that the Spanish court had ordered carronades, 96, 69 and 42 pounders, for the lower gun deck, upper deck and quarter deck, respectively, of a vessel of 74 guns. Gascoigne had also approached the French court, where his promotion material had been graciously approved.[24]

As yet, this is all I have been able to find about the delivery of carronades to the Dutch fleet before 1795, i.e. before the advent of the Batavian Republic with its generally pro-French attitude. We know that the Batavian navy consistently issued carronades to their ships, on a larger scale than previously and that they developed their own patterns.

However, the number of carronades used by the Dutch during the battle off Camperdown in 1797 was pitifully small and carronades were, in fact, found only on one or two of the lighter vessels.

Carronades continued to be used in the Dutch navy until well into the second half of the nineteenth century. In the 1860s, the Dutch also developed a rifled carronade. This was a 60-pounder carronade with trunnions, the bore of which was cast full with bronze and then drilled out with rifling to take shells of 16 cm (about $6\frac{1}{3}$ inches).

The Dutch army adopted the carronade in the nineteenth century and used it on rowing launches to defend the moats around fortifications.

Notes

1 All figures regarding the shipbuilding programme of the Dutch navy from the late 1770s onwards are taken from: J C de Jonge, *Geschiedenis van het Nederlandsche Zeewezen*, Vol. 5, 2nd edition, Haarlem, 1862.

2 De Jonge, Vol. V, *op cit.*, p. III.

3 De Jonge, Vol. V, *op. cit.*, p. 8–9.

4 On Van Kinsbergen, see: R Prud'homme van Reine, *Jan Hendrik van Kinsbergen 1735–1819: admiraal en filantroop*. Doctoral diss. Univ. of Leyden; in progress.

5 *Beginzelen der practyk in de artillerie, voor de adelborsten op 's Lands oorlogscheepen* ('Principles of practical artillery, for the midshipmen on the warships of the nation') Amsterdam, 1782, p. 11.

6 De Jonge, Vol. V, *op. cit.*, p. 57.

7 F L Robertson, *The evolution of naval armament*, (reprint of the 1921 ed. London, 1968, p. 127). On the Carron Company, *cf.* R H Campbell, *Carron Company* Edinburgh, 1961.

8 Robertson, *op. cit.*, p. 132.

9 De Jonge, Vol. V., *op. cit.*, p. 12 note 1.

10 The records of the company are in the custody of the S.R.O. under the reference GD.58. Most of the company's early records were deposited with the Scottish Record Office in the 1950s and, more recently, further records were acquired from the liquidator appointed to the company. Information kindly supplied by Peter G. Vasey, Esq, West Search Room, Scottish Records Office, on 13 May 1986.

11 S.R.O. GD.58/6/1/17 p. 158 dated 27 August 1779. Information supplied by Mr Vasey on 14 July 1986.

12 De Jonge, Vol. V, *op. cit.*, p. 12. This report, along with some other interesting manuscripts regarding carronades, is kept in the Royal Netherlands Household Archives, in the correspondence between the Stadtholder Prince Willem V and Van Kinsbergen, Inv. No. A31–261 III.

13 A.R.A., le Afd., Admiraliteit/XXXIX/26.

14 A.R.A., le Afd., Admir./XXXIX/26: Contract dated 29 November 1783 between Charles Gascoigne and Ten Cate, De la Lande and Fijnje. Information kindly supplied by Drs J R Verbeek of The Hague, in September 1987.

15 A.R.A., le Afd., Admir./XXXIX/26: Letter dated 17 May 1783 from Gascoigne to Ten Cate. Information supplied by Mr Verbeek in September 1987.

16 A.R.A., le Afd., Admir./XXXIX/25: Memorie omtrent de lengte en swaarte van het geschut voor oorlogsschepen: rec. Jan. 1784 van den Capitein Grave van Byland ('Memorial on the length and weight of the artillery for warships . . .'). The memorial says literally (in translation): '. . . the cheeks of which tore apart by the violence with which the pieces recoiled'.

17 In the Netherlands Royal Household Archives, the *Koninklijk Huisarchief*, in The Hague there is yet another interesting file on carronades. That was put together in 1785 by the Dutch envoy in London, D W baron van Lijnden, and sent by him to the Stadtholder on 6 May. This file contains a report written in French by a J J Jorez on the carronade in general, recommending it in the highest terms, and giving many examples, measurements and comparisons. Two adjacent letters, written in the same hand, are: (Copy) Letter from Mr James Baird, dated Sheerness Aug.¹ 2 782, concerning the experiments made with the carronades on board the Rainbow, laying of [sic] Sheerness, at the mouth of the Thames, and an Extract of a Letter from General Melvill (dated London May 15, 1781), in which several trials are reported. The file further contains a printed broadside entitled Experiments made at Carron river-mouth, the 4th of September 1781, with 100 pounder carronade of 6 diameters, length of the bore (powder-chamber included) 4 feet $9\frac{1}{2}$ inches, diameter of the bore $9\frac{1}{8}$ inches, diameter of the chamber 8 inches, weight about 50 Cwt. A hand-written note at the bottom states: *N.B. There was cast a 132 pounder carronade in January 1782. Experiments were made in the above manner.*
Royal Netherlands Household Archives, Inventory of Prince Willem V, ref. A18–291 III.

18 A.R.A., le Afd., Admir./XXXIX/25: Remarques wegens het gebruik van carronades ('Remarks on the use of carronades'), dated 25 August 1784, probably by W. Gerlach.

19 A.R.A., le Afd., Admir./XXXIX/26: Letter dated 9 February 1784 by Charles Gascoigne to Van der Hoop.

20 The earliest dateable illustration of any carronade found in Dutch records is the rather well-known print of a 'Joint carronade fitted to the ships [sic] side. Plate 1st'. It is among the Van Lijnden file to the Stadtholder (Royal Netherlands Household Archives, Inv. No. A18–291 III). This print can therefore conveniently be dated at no later than May 1785.

21 A.R.A., le Afd., Admir./XXXIX/26: Letter dated 9th February 1784 by Charles Gascoigne to Van der Hoop.

22 A.R.A., le Afd., Admir./XXXIX/26: Report dated 23 Jan. 1779 by John Zuil to Charles Gascoigne.

23 A.R.A., le Afd., Admir./XXXIX/26: Letter dated 9 February 1784 by Charles Gascoigne to Van der Hoop.

24 A.R.A., le Afd., Admir./XXXIX/26: Letter dated 29 September 1783 by Charles Gascoigne to Ten Cate. Since this article was written, it has become known that the Carron Company did supply many guns to the Dutch East India Company in 1789 and 1790. It is, however, not yet known at this stage of the research if there were any carronades among them. Information by Mr Vasey on 30 November 1987.

Gunlocks: Their Introduction to the Navy

J M BINGEMAN

The recovery of over 2,000 gunflints of unusually large size from the wreck site of *Invincible*, a 74 gun mid-eighteenth century vessel, prompted my investigation into why she was carrying so many flints. Most flints were of two sizes, 37 mm and 44 mm at the striking edge, and were unusual because they were far larger than those used for muskets. However, after enquiries at the Museum of Artillery at the Rotunda, and discussions with Seymour de Lotbiniere, author of *English gun flint making in the 17th and 18th centuries*, no explanation can be found as to why *Invincible's* wedge-type flints were so large (Figure 1).

If you read through the admiralty 'In/out' letters of this period, there are various refer-ences to authorised experiments in the fleet for both locks and tubes to improve the accuracy of gunnery afloat. One particular letter dated 21 October 1755, named thirteen ships including *Invincible*, with the instruction *quarterdeck guns of all his Majesty's ships now fitting and refitting for sea to be fitted with locks, and the gunners may be supplied with a sufficient number of tin tubes for priming them.*[1] Less than two years later, a minute from the Board of Ordnance, dated 17 August 1757, notes the decision to fit cannon locks for all quarterdeck guns of his Majesty's ships in commission. The suggestion that the flints for these locks were large came from a further Board of Ordnance minute, dated 20 October. It suggests that any deficiency in the 'cannon' flints available could be made up by selection from the *common musket and carbine flints* already in store.

We do not know whether the trial locks fitted to the *Invincible's* 9-pounder quarterdeck guns will be found. The Court Martial records that her quarterdeck guns were jettisoned when she first went aground.[2] I have not been able to find any record of their recovery. However, there are various magnetometer signals to the south that may be caused by these cannon.

The consistency of the 2000 flints recovered suggests that they originated in Kent. It is probable that they were manufactured by William Levett of Northfleet, the principle supplier of the Board's gunflints between 1742 and 1781. Flints cost 4s 6d per thousand, although larger 'wallpiece' flints were more than three and a half times the price of musket flints.[3] The 2000 flints represent a loss of £1 11s 6d to the Crown.

In France, a proposal by the Master Armourer at the Port of Toulon, suggested

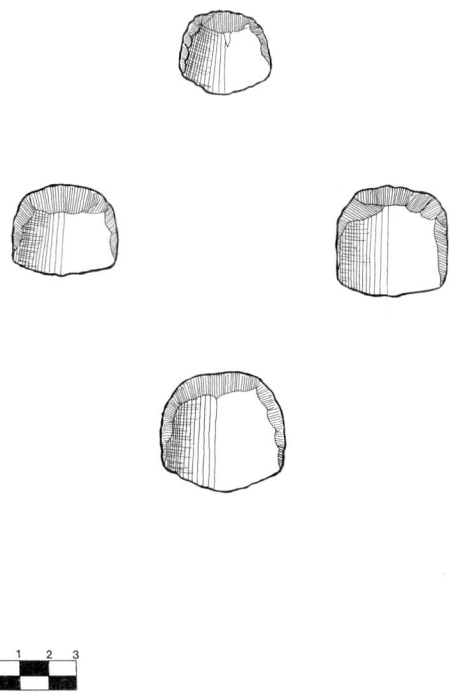

0 1 2 3
CM

Figure 1   Four sizes of flints from the *Invincible* (1758) Wreck site.

Figure 2   A 10 × 12 inch apron of lead from the *Invincible* (1758) Wreck site.

that firing locks tried on sea-service guns as early as 1728.[4] However, this is predated by a letter from the Principle Officers of His Majesty's Ordnance, dated 20 July 1745, which reports unfavourably on lock trials and also refers to a similar fitting of locks to mortars some 40 years earlier. *They were all condemned and sold being found quite unfit for service as I apprehend you will find this new invention will prove for firing of cannon.*[5] This shows that there is some evidence of cannon locks going back to the first decade of the eighteenth century. Admiral Steuarts writing to the Admiralty on 2 October 1747, stated that: *Captain Harland is the only one at present here to whom I gave orders about the new invented locks for cannon, and from him I am in daily expectation of knowing how far they are found useful.*[6] Not unexpectedly, there was considerable opposition to their introduction in the fleet. There were complaints of the flints getting wet and failing to fire. On the subject of tin tubes, introduced at the same

time to transmit the spark from the lock pan down the vent hole, Vice Admiral Hawke wrote following the Battle of Quiberon Bay on the 20 November 1759: [*they are*] *pernicious things apt to fly out and wound the men.* Two months previously, on 15 September, while on board the *Ramillies*, he had written to John Cleveland, Secretary to the Admiralty Board. *Endorsed are seven opinions concerning cannon locks and tin tubes in mine, both of them will be very useful.* Interestingly, a flint from the *Ramillies* wreck (1760) was identical to the 2000 found in the *Invincible*.

The practice had been to cover the vent hole with an 'apron of lead', to keep the vent dry. This was no longer possible with the ew locks and explains why the flints were getting wet. Two sizes of lead aprons have been recovered from *Invincible*: 10 by 12 inches; and $5\frac{1}{2}$ inches square, and these would have been useless once the locks were fitted (Figures 2 and 3). The 10 × 12 inch apron size agrees exactly with Blackmore's

Figure 3 A $5\frac{1}{2} \times 5\frac{1}{2}$ inch apron of lead from the *Invincible* (1758) Wreck site.

specification, while the $5\frac{1}{2}$ inch square size is a slight variation on Blackmore's $4\frac{1}{2} \times 6$ inch size.[7] Sometime after this, a modified lead apron with a 'hump' was introduced to overcome the problem of keeping the firing mechanism dry (Figure 4). Exactly when this happened is unknown, but examples were found in the wreck of the *Pomone* (1811) of 18 lb guns and 32 lb carronades.

The commissioners executing the office of Lord High Admiral, personally issued the precise instructions to standardise gun drill. Woe betide any captain who failed to exercise his gun crews daily. Gunlocks certainly contributed to the supremacy of British gunnery in one important detail. They gave the captain of the gun complete control of the moment of firing as he pulled the lock lanyard, rather than by verbally ordering his no. 2 to fire with the traditional match (Figure 5). As we said, all gun drill was standardised throughout the fleet. Drills were laid down for a variety of manning levels, from a full crew of thirteen to a minimum crew of six. To ensure that maximum fire rates were achieved, the crew went to quarters for gun drill every morning and the varying manning levels were exercised. It was therefore not by chance that

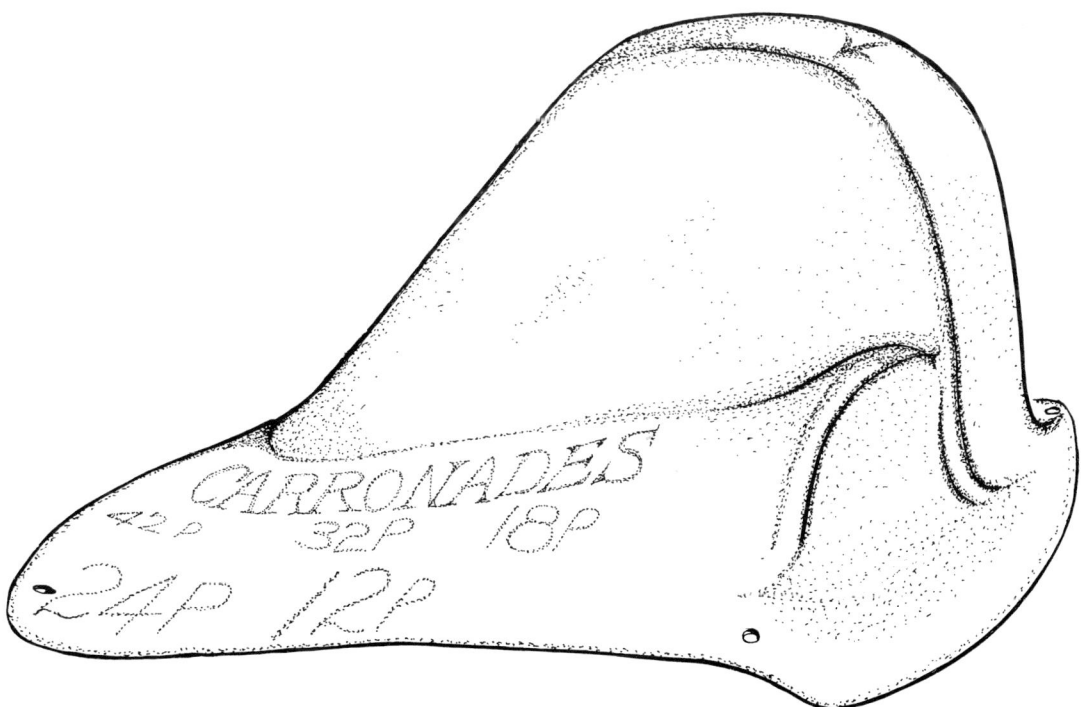

Figure 4 The new style apron of lead to cover a flint-lock. Designed to fit a number of calibres; this one had been fitted to a 32 lb carronade and was recovered from the *Pomone* (1811) Wreck site.

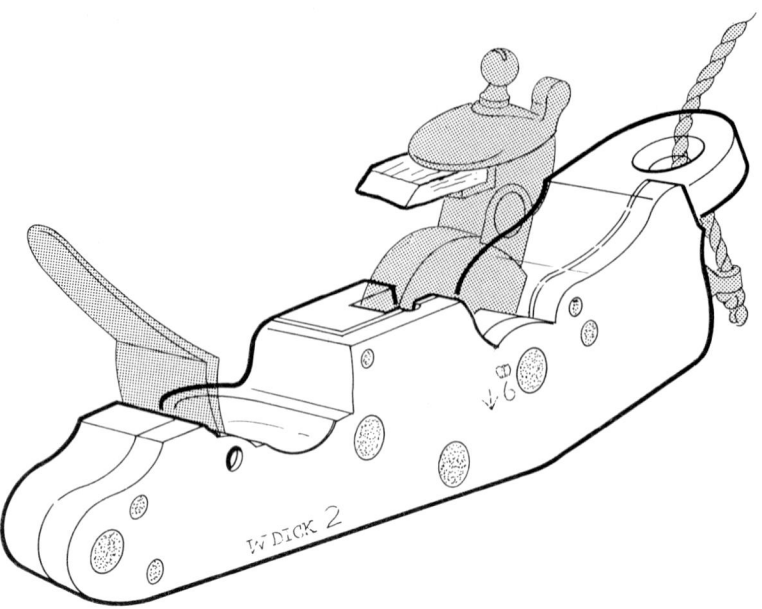
Figure 5   A flint-lock from the *Pomone* (1811) Wreck site.

British naval gunnery remained pre-eminent among the World's navies, but the painstaking attention to every detail that made the English the undisputed masters of the seas in the eighteenth century.

Acknowledgement
The author wishes to acknowledge the help of Mr A T Mack who carried out much of the research for this paper.

Notes
1 PRO Adm 2/219.
2 PRO Adm 1/52973.
3 WO 50/19:20, WO 48/81:379.
4 Boudriot's letter to Mack dated 11 January 1985.

5 PRO Adm 1/4008.
6 PRO Adm 1/916.
7 H L Blackmore, *The Armouries of the Tower of London*, London, 1976, p. 218.

The Restoration of the 32 pounders at Fort Nelson

A C CARPENTER

Five cast-iron smooth-bored guns, still mounted on their original iron carriages and slides, have been displayed at Edinburgh Castle since the beginning of this century. Two further guns, also with their carriages and slides, were stored at Fort George near Inverness, making a total of seven pieces.

Their official designation, as printed in the 1898 Handbook was:

32 — PR. S.B. B.L. gun (for flank defence). Mounted on garrison sliding carriage and traversing slide. Land service 1898.

In 1838 a Mr Monk, one of the staff of the Woolwich Arsenal, involved in administration and design, received approval to cast a number of iron guns which he had designed, ranging from 32 to 56 pounders. The 32 pounders were produced in three sizes referred to as Monk Patt. A, B, and C:

50 cwt 9 ft 0 in 6.37 in Bore Monk Patt. A.
45 cwt 8 ft 6 in 6.35 in Bore Monk Patt. B.
42 cwt 8 ft 0 in 6.35 in Bore Monk Patt. C.

These were issued to the British Naval Service in the late 1840s.

In the 1860s and 70s various programmes were set up to re-use the best of the old cast-iron smooth-bored guns still on government registers. For example some cast-iron shell guns by Millar and Dundas were converted to rifled pieces on the Palliser System.[2]

Another programme initiated by Woolwich Arsenal converted the cast-iron Monk Patt. C. guns of 42 cwt which were, of course, muzzle loading, to breech-loaders.

This was achieved firstly by removing the whole of the cascabel area of the guns up to the breech ring. This still left an area of solid cast iron at the breech face end of the bore (Figure 1). This was then bored out, making the gun effectively a tube open at both ends. The open end of the breech was then relieved to a set diameter, which still allowed for an interrupted thread to be cut in the wall of the breech. This thread would receive its counterpart cut around a solid steel breech block. The shoulder of the bore of the gun was then machined to receive a copper ring to obturate the breech by the use of an Elswick Cup.[3] With the removal of all this metal from the breech of the gun, there still remained the original copper-lined vent used to fire the gun

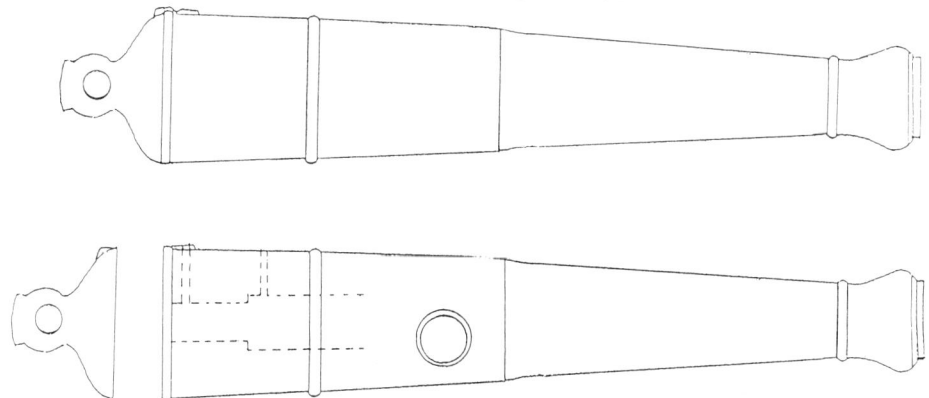

Figure 1   Cast iron 42 cwt gun, Monk Pattern 'C', showing area of breech removed.

in its original muzzle loading state. This vent was machined out and sealed with a threaded spigot and a new vent was drilled forward of the old position to meet the re-positioned powder chamber. As with the Palliser conversions, the date of conversion was chiselled on the trunnion face (Figure 2).

In 1984, the seven guns, together with their carriages and slides, from Edinburgh Castle and Fort George, were moved from Edinburgh Castle to Fort Nelson, Portsmouth. These guns are Monk Patt. C. of 42 cwt converted to breech loaders. All seven guns had, however, lost their breech blocks.

It was following their removal to Fort Nelson that I was approached by the Royal Armouries with a view to having the missing breech blocks and carriers made and fitted to these guns.

Detailed information regarding the breech conversion was very scarce and all that could be found was in a drawing in the *Treatise on Service Ordnance*.[4] Further information was also gleaned from an excellent paper by Baker.[5]

The engineering firm of Fox and Haggart in Plymouth had, over the past thirty years, carried out various contracts for the Department of the Environment in relation to period armament and were engaged to undertake this project.

All seven guns were transported to Plymouth and, on arrival, the piece with the best interrupted thread was inspected and thoroughly cleaned. The next procedure was to check the number of threads per inch and their pitch within the gun's breech.

Although modern screw-cutting lathes are capable of cutting a large number of different threads it was impossible to reproduce the exact screw thread cut into the breeches of the guns. However, an arrangement of various odd gear wheels was set up on a lathe to cut, and match as near as was possible, the thread needed. It was then found necessary to hand file and fit the threads to ensure that the breechblock reached its correct position within the chamber.

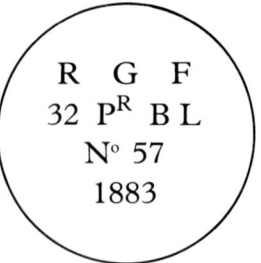

Figure 2   Trunnion mark showing that the conversion was carried out at the Royal Gun Factory at Woolwich.

Each breechblock, with its three interruptions to the thread, was made, machined and fitted individually, none of them being interchangeable within any of the seven pieces. Considerable time and effort went into the fitting of the lock and bell-crank levers (Figure 3).

The design for the cup obturator, called the Elswick Cup, was taken from *Naval Gunnery* (Figure 4).[3] The obturation of the breech in the Elswick Cup System was by means of a steel cup, designed to open out by the rearward pressure on the cup face at the time of the explosion, and seal against a copper annular liner inside the breech wall. It was a system which required the breech to be cleaned regularly to maintain gas-tight operation, and was succeeded by the efficient de Bange Pad.

The details for the manufacture of the breech block carrier were found in the *Treatise of Service Ordnance*.[4] Figure 5 gives a general indication of the design and the various components required to make the breechblock, carrier and lever etc. In the original drawing the gun metal parts were coloured yellow; the carrier and lever for the breechblock.

In order to manufacture the components a steel copy was made of the carrier and breech lever, the whole was set upon the gun's breech face and adjusted so that it was fully functional. When this was completed the steel patterns were sent to a pattern maker for him to produce wooden examples to the specified

Figure 3　A complete breech block, lever and carrier showing the Elswick Cup Obturator and locking components.

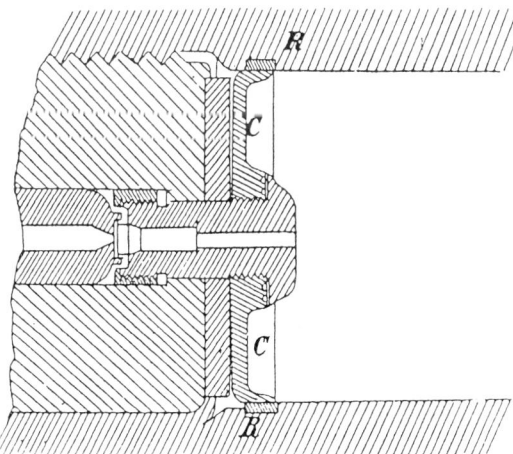

Figure 4　Section through the Elswick Cup at the face of the breech block.

dimensions allowing for shrinkage when produced in the non-ferrous metal specified — gun metal.

The new gun metal castings had to be machined to accommodate the various levers, springs, bell-cranks etc. All these items were required to operate the carrier and block and to allow the breechblock to travel in and out for opening and closing the breech. Springs of lever type were made by the author from spring steel and annealed and tempered to activate the locking components.

With all the components for one gun completed and its breech mechanism set up and working quite satisfactorily, it was then a matter of manufacturing all the components required to finish the seven pieces (Figure 6). These were transported back to Fort Nelson for mounting on their respective carriages. Each breechblock and its components was given a number to correspond with that of each gun on the breech face, so that components were not mixed up at fitting or dismantling.

The gun was designed to operate within a fortification and one of its roles was to cover the flanks of ditches from caponiers within the ditch, firing canister and grape shot which, being of low velocity, was suited to a cast-iron

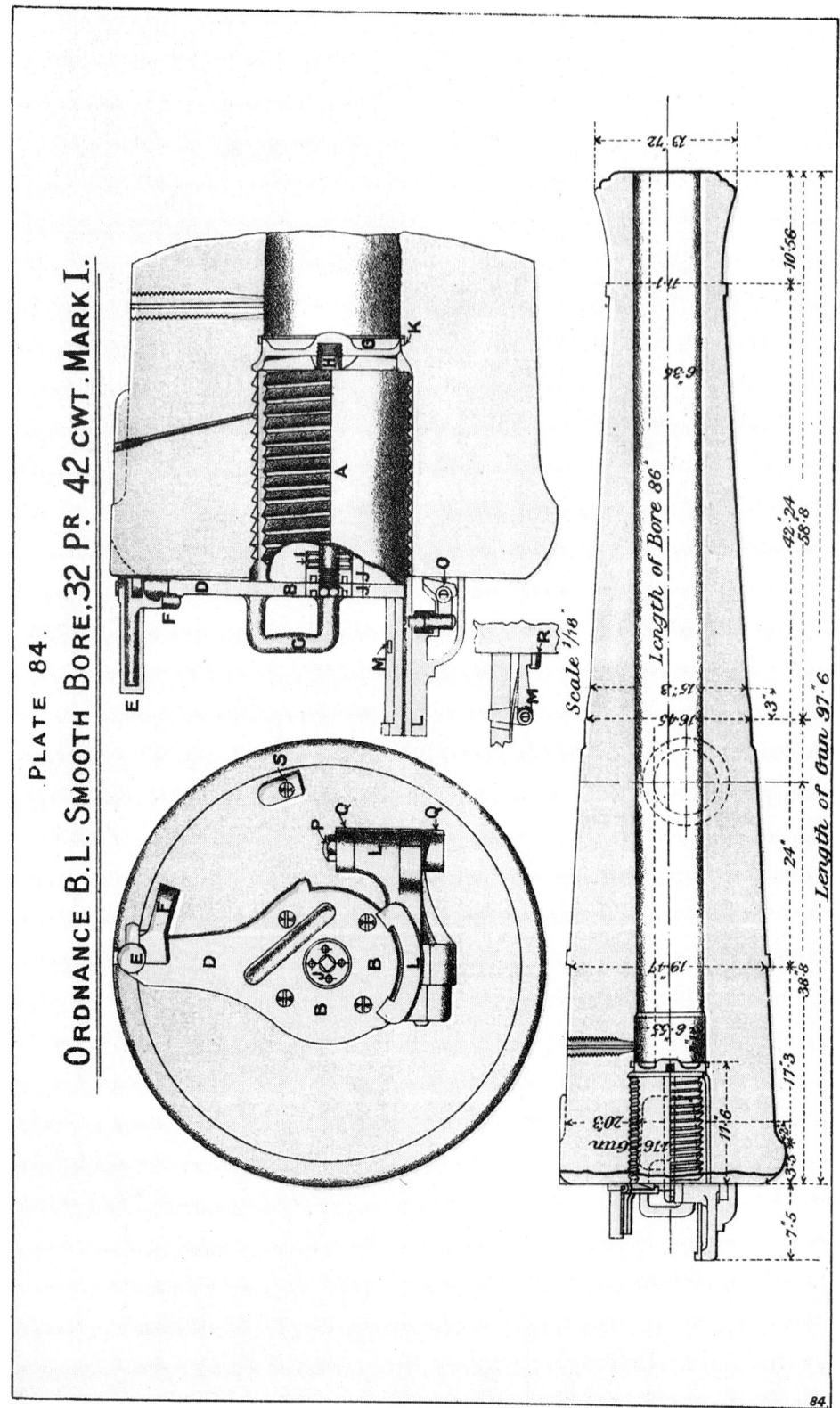

PLATE 84.

ORDNANCE B.L. SMOOTH BORE, 32 PR. 42 CWT. MARK I.

Scale 1/16"

Figure 5   Breech mechanism, in section, used to obtain manufacturing detail.

Weller & Graham L.ᵈ Litho. London.

Figure 6   Four of the guns with the newly made and fitted breeches.

gun with a breech-loading system using cast-iron threads.

The carriage and slides of iron were especially designed for this converted gun and had an overall length of 6 ft 8 in with a rearward slope uphill of 10°, to rapidly absorb the recoil of the carriage and gun working in a confined casemate (Figure 7). The carriage accommodated a hand-wheel which elevated a screw working under the tongue of a stool bed on which a wooden quoin was laid supporting the gun breech. The slide was set upon iron rails to allow the whole to be traversed to left or right.

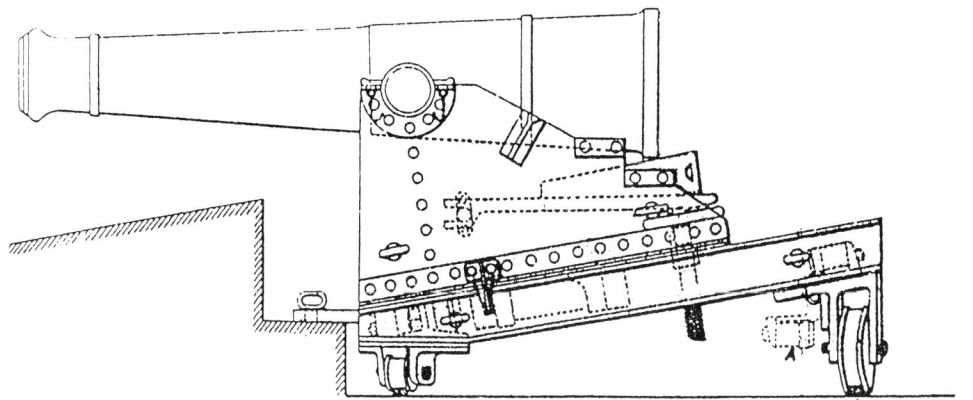

A. Rear buffer stop, folded down for housing.

Figure 7   Cannon on its carriage as set up in caponier casemate.

As with most old ordnance on display, nearly all moveable fittings had been removed. This made it necessary, in the case of these seven guns, to renew the front and rear trucks to the slides, the elevating screws, hand-wheels and wooden quoins to the carriages as well as various bolts and pins.

In the autumn of 1986 the author fired a number of blank charges, by friction tube ignition, ranging from 8 oz to $2\frac{1}{4}$ lb of black pebble powder with junk wads, to see if obturation was effective. This was not expected owing to the corrosion in the internal breech areas and the pitted copper breech liners, which had been exposed to the elements for something like 80 years. As photographs show gas coming from the breech area, our assumption that obturation was not effectively obtained was substantiated.

The service charge for the Monk Pattern C converted gun was 3 lb of black powder to fire canister shot.

References

1 Douglas, Gen. Sir H, *A Treatise on Naval Gunnery*, 1855.
2 *Treatise on the Manufacture of Guns*. HMSO. 1886.
3 Garbett, Capt H, R.N., *Naval Gunnery*, 1897.
4 *Treatise on Service Ordnance*, Plates, 1908, HMSO.
5 Baker, H A, 'Late Cast Iron Service Guns,' *Journal of the Arms and Armour Society*, Vol. IX, No. 3, 1978.

The Development of Naval Guns 1850–1900

N J M CAMPBELL

Progress in the development of naval guns was far greater during this period, than in any previous century, and many items have been omitted in the interests of space. Attention has been concentrated on the larger weapons, where construction and attainment of the desired performance are most difficult.

Although a few rifled guns were mounted in ships during the 1850s, most of the armament was still cast-iron smooth-bores. For example, the 'solid-shot' long 32 pounder guns of 58 cwt, the last of this famous calibre, and the light 68 pounders of 95 cwt — the best of all the cast-iron smooth-bores. A new 9 ft 6 in 42 pounder gun of 67 cwt was tried, but not introduced, as an alternative to the 32 pounder.

Of the heavy-chambered 'shell' guns, a slightly modified 10 in of 86 cwt was introduced in Britain. The only major development in this type of gun was in the United States, where 11 in Dahlgren guns were adopted. These guns had conical chambers and a distinctive pear-shaped external form, without the usual raised bands or rings (Table 1).

The Dahlgren gun also used 20 lb 'distant' charges. They could fire solid shot, weighing 166 lb. Ranges with standard charges and shell were about 1700 yards at 5° elevation and 2600–2800 yards at 10° elevation. The 68 and 32 pounder guns also fired shell, the weights being 51 and 24 lb respectively. Ranges for these two guns, firing shot with the charges given in table 1, were 1900–2000 yards at 5° elevation and 2800–2900 yards at 10°. It was unusual for the centre of gravity of cast-iron shot to be at the geometric centre, and ranges could be increased by loading shot with the CG off centre. In June 1852, an experimental 10 in/116 cwt gun firing a 100 lb shell, made deliberately eccentric, with a 16 lb charge, achieved a maximum range of 5860 yards at 32° elevation.

Some interesting, experimental, wrought-iron smooth-bores were made, the most remarkable being the Horsfall gun and the Mallet mortar.

The Horsfall gun was a larger version of the guns mounted in the *USS Princeton* and resembled these guns as they were originally built, made from a forged mass of wrought iron and not hooped. The gun weighed 24 tons, of 13 in bore and was 12.3 calibres long. The solid cast-iron shot was 280 lb with a windage of only 0.2 in — the normal charge being 50 lb, although up to an 80 lb charge could be used. During trials in November 1856, a 50 lb charge reached a range of 5000 yards at 18° elevation. The high-air resistance of large spherical projectiles was shown by the 8 in Lancaster rifle, which ranged 300 yards more at the same elevation. Incidentally, the Horsfall shot fired point-blank, stopped 5346 yards from the gun. In 1862 a muzzle velocity of 1631 ft/sec was achieved with a 74 lb charge. The vent was bushed after 70 rounds, which compares well with 250 rounds for a 32 pounder. Such a large gun was perhaps an anomaly for the time. As an armour piercer, it would have compared well with the 15 in smooth-bores used so much in the American Civil War.

The Mallet mortar was originally intended for the siege of Sebastopol, but the first of the two guns produced was not finished till 1857. With a bore diameter of 36 in the mortar weighed 42 tons, and could be broken down to components of 12 tons maximum. The barrel consisted of wrought-iron rings, held between a heavy muzzle piece and a cast-iron base by

Table 1

Gun	*Length		Bore	Windage	Length of bore	Weight of gun	Nominal weight of projectile	Charge	Muzzle velocity
	ft	in	in	in	calibre	cwt	lb	lb	ft/sec
68 pdr/95 cwt	10	0	8.12	0.198	14.0	95	68 shot	16	c1579
32 pdr/58 cwt	9	6	6.375	0.198	17.0	58	32 shot	10	1690
11 in Dahlgren	—	—	11.0	0.150	12.0	140	133 shell		

*Length = muzzle to rear of base ring.

longitudinal bolts. The shells weighed from 2360 lb to 2940 lb empty, with a burster of about 480 lb planned. The greatest propellant charge used in trials with one mortar was 80 lb. However a number of minor failures occurred, and in 1858/9 unfortunate parsimony over a few hundred pounds relegated both mortars to museum pieces.

The 1850s saw the introduction of rifled guns, previously known, but generally neglected for three or four centuries. Rifled guns had many advantages: a heavier shell from a gun of given size; greater range due to reduced air-resistance from an elongated projectile, even if the greater weight reduced muzzle velocity for a given charge; and greater accuracy.

The Cavalli gun of 1846 was a 6.5 in cast-iron breech-loader, 14.3 calibre in length, weighing 66 cwt. Windage was not prevented, rifling being by $2\frac{1}{4}$ in deep ribs cast on the projectile. These engaged with two helical grooves in the bore. The breech was closed by a cast-iron plug, a copper ring on the rear face and a horizontal, hardened wedge which passed through the body of the gun.

The Wahrendorff gun, also of 1846, was similar. It also had a 6.37 in bore, but had a different breech. The breech plug was attached by a rod to a hinged door. A horizontal wedge was slotted to pass over the rod. On closing the breech, the door was swung into position, the plug pushed forward and the wedge driven in. The whole was tightened by

a nut on the end of the rod. In 1850, both these guns were tried against a 32 pounder/ 56 cwt smooth bore.

With a 64 lb shot and 10 lb charges, the two rifled guns ranged about 1100 yards more at 15° elevation, although the deflection was very variable. The Wahrendorff gun withstood the test well, but the Cavalli gun blew its breech off after a few rounds. It should be mentioned that breech-loading had been a neglected principle, like rifling, although breech-loading is even older in origin.

Although it was not officially adopted, the Lancaster muzzle-loading rifled gun was used in the Crimean War. It was first tried in 1851, when an elliptical helical bore of about 8 in was used, the diametrical difference being 0.27 in. The projectile was also elliptical. The guns were mostly bored from 68 pounder/ 95 cwt or 8 in/65 cwt castings, and the shells used in the Crimean War were up to about 100 lb in weight. This rifling system was not a success, as the projectile was liable to jam and burst the gun, or to break-up in the bore. The range was adequate, as we saw in the comparison with the Horsfall gun, but at both Sebastopol and Bomarsund, the guns were found to be inaccurate. However, at Sveaborg in 1855, a Russian ship, the 3-decker *Rossia*, was twice set on fire by gunboats armed with these guns.

Rifled gun development had also progressed in France. The first gun adopted by the

French Navy was the 6.4 in muzzle-loader model of 1855. Tests on this gun were held to have made wooden battleships obsolete. The gun was cast to the outside form of the 8.8 in shell gun of 1827–41. Few were made. Two lugs, cast on the projectile, engaged in the rifling grooves. This system was liable to jams and burst guns, but was improved by the use of zinc lugs. The shell weighed 58 lb and a charge of $7\frac{3}{4}$ lb was used. Up to 1300 yards the accuracy was similar to that of the 32 pounder smooth-bore, but at longer ranges it was much better. The rifled gun was said to shoot accurately at 4400 yards.

An important invention of this period was the so-called 'French' breech with an interrupted screw block, which was subsequently widely used. This was not in fact French, being first patented in 1853 in the United States, by Schenkl and Saroni.

There was certainly a big difference between various types of armoured ship between 1858 and 1865. Indeed it is hard to imagine two more different ships of similar date than the *Magenta* and *Dictator*. This difference was mirrored in the armour-piercing guns of the period. Smooth-bores, muzzle-loaders and breech-loaders, rifled on very different systems, were all favoured. In their construction, cast iron, wrought iron and steel were all tried in various combinations — cast iron with steel or wrought iron, or wrought iron with steel. In France, the 6.4 in RML of 1855 was succeeded by the model of 1858, cast to the form of the 40 pounder of 1786. The first guns were made completely of cast iron, but in November 1859, hoops, shrunk over the breech, were adopted — one layer of puddled steel hoops being used. Hoops were also applied to those guns already cast. The rifling consisted of three rather deep grooves with increasing twist. The projectile had zinc studs which engaged with the grooves. The 6.4 in RML of 1860 differed only in the form of the rifling groove. The 6.4 in BL of 1860, which used an interrupted screw breechblock with obturation by a steel diaphragm, was also

of the same dimensions and performance. These guns were intended to fire cast-iron shells of about 67–70 lb with propellant charges of $7\frac{3}{4}$ lb. As this was completely useless against armour, each gun was allowed to fire 20 solid steel cylindrical shots of 99 lb with a $16\frac{1}{2}$ lb charge. This meant the possibility, though not a very probable one, of piercing $4\frac{1}{2}$ in wrought iron at 100 yards (Table 2).

It was obvious, however, that these 6.4 in guns were not sufficient. In 1861, trials were made of an all steel 6.4 in hooped BL with a bore 28 calibres long, firing a 99 lb shot with a charge of $26\frac{1}{2}$ lb, giving a muzzle-velocity of 1350 ft/sec. Tests against armour were not considered satisfactory, and, in spite of the desire to retain relatively small calibres, it was clear that larger-bore guns were necessary. These were introduced in the models of 1864, sizes being 6.4 in, 7.6 in, 9.4 in and 10.8 in. All were cast iron and steel-hooped, with two layers of hoops in the three largest guns. They were BLs with the interrupted screw breech, rifled in a similar war to the previous guns. The three main differences were that from 7.6 in upwards there were five grooves, most of the projectiles had bronze rifling studs, and the powder chamber was about $\frac{1}{2}$ in larger in diameter than the bore. The most important of these made for ships, completed or being built in 1865, were the 7.6 in and 9.4 in (Table 2). Muzzle velocities were too low with the type of powder in use at that date, not exceeding 1135 ft/sec, for the guns to be satisfactory. Rifling was also not ideal, the localised load being high. The appreciable windage (0.120 in in the 9.4 in) led to gas escaping past the projectile which scored the barrel.

Apart from the Lancaster oval-bore, tried in the Crimean War, the first British rifled gun to be adopted by the services was the Armstrong BL. This was rifled with a large number of small grooves, 76 in the 7 in, of uniform twist. The projectile was lead covered, the diameter over the coating larger than the bore, and the rifling grooves thus cutting into the lead. The

Table 2

Gun	Bore	Length of bore	Weight of gun	Nominal weight of projectile	Charge	Muzzle velocity
	in	*calibre*	*ton*	*lb*	*lb*	*ft/sec*
French 6.4 in (1858–1860)	6.48	16.7	3.6	99	16.5	1055
French 9.4 in (1864)	9.45	17.4	14.3	318	53	1120
French 7.6 in (1864)	7.64	18.0	7.8	165	27.5	1135
Armstrong 7 in BL	7	14.2	4.1	110	14	1175
100 pounder SB	9	11.7	6.25	102	25	1630
150 pounder SB	10.5	11.9	12	168	40	1630
7 in RML	7	15.8	6.5	115	22	1430
8 in RML	8	14.7	9	180	30	1363
9 in RML	9	13.9	12	250	43	1336
15 in SB	15	9.7	19.2	452	60	1220

breechblock, inserted radially, had a coned face and also served as the vent-piece, by which name it was known. This was kept in position by an axial hollow breech-screw, through which the gun could be loaded, once the vent-piece was withdrawn. Originally the gun was to have a steel barrel, supported by shrunk-on wrought-iron coils. However, very few guns were made with steel barrels. Some had wrought-iron forgings, while most of the larger guns had coiled wrought-iron barrels. The first gun, to fire a 3 lb projectile, was tried in July 1856, and proved successful. An 18-pounder gun was tried successfully for army use, and the Armstrong system was adopted in January 1859. In that year a 40-pounder gun of 4.75 in bore was introduced for the navy, but it served little purpose against armoured ships. In September 1859, a 6 in gun firing a shot of about 80 lb with a 12 lb charge, was tried against the brittle 4 in plates of the floating battery *Trusty*. Two steel shot managed to pierce it, one at 200 yards and the other at 400 yards, but cast-iron shot failed. In 1860, during further trials using a 10 lb charge against good quality rolled $4\frac{1}{2}$ in plates, little damage was done. However, shortly after the *Trusty* experiment, a 7 in gun of 72 cwt was recommended as the replacement for the 68 pounder SB. Unfortunately the recoil was found to be too violent and only 76 were

made. A heavier 7 in of 82 cwt was introduced in 1861 (Table 2). Its armour-piercing performances was disappointing. With iron shot, it was considered inferior to the 68 pounder, though with steel shot of good quality, unbacked $5\frac{1}{2}$ in plate was pierced at 200 yards. In addition, neither the system of rifling, nor the coiled barrels were satisfactory, and the vent-pieces gave much trouble. The BL guns did not behave well in action, for example in Kagoshima in August 1863, and shortly afterwards they were withdrawn from service.

Several other rifled guns were considered before the Armstrong was finally adopted. Of these, the Whitworth system merits description. The bore was a spiral hexagon with rounded corners and the projectile of similar shape. Both BL and ML guns were made. Construction was by a mild steel barrel with mild steel hoops forced on cold by a hydraulic press. Whitworth guns have a number of successes to their credit. In October 1858, a 68 lb wrought-iron shot fired with a 12 lb charge from a 5 in/5.5 in ML, easily pierced a 4 in wrought-iron plate and the wooden side of the target ship at 450 yards. This gun was bored from a standard 68-pounder casting and shortly afterwards it burst. Its muzzle velocity was about 1280 ft/sec. Nearly four years later, in September 1862, a built-up ML of similar calibre sent a $68\frac{1}{2}$ lb shell through a small target plate, 4 in thick, at 200 yards. Apparently this is the first instance of a shell piercing 4 in plates. The shell was steel, flat-ended and unfuzed. Impact caused sufficient friction in the charge to burst behind the plate. The best performance was in November 1862 when a wrought-iron 6.4 in/7 in ML sent a 151 lb shell with a 5 lb burster through a very good quality 5 in plate at 800 yards. The shell, of similar type to the smaller one above, burst in the backing. This gun weighed 149 cwt and had a barrel 126 in long with a rifling twist of 1 in 20. Windage was 0.04 in and a 27 lb charge gave a muzzle velocity of about 1370 ft/sec. For its period this was a powerful gun. In comparison the 7 in Armstrong BL, with 110 lb projectile and battering charge of only 14 lb, had a muzzle velocity of 1175 ft/sec.

The success in armour-piercing was due to both good quality projectiles of relatively large weight/cross-section ratio as well as a gun sufficiently strong to withstand the charges necessary for adequate velocity. The system of rifling was incidental to the gun's performance. It was not liked because it was expensive to produce and liable to jam, having two similar metal surfaces working together at high speed. The particular gun used became, in fact, unserviceable after only 19 rounds due to a large flaw in the bore and excessive wear.

Armstrong also developed guns with his polygroove system of rifling and a wedge breech, of which a few 6.4 in guns were made for land service. Rifled muzzle-loaders also used 'shunt' rifling. In this system the depth and width of the grooves varied in different areas, so that a studded projectile could be loaded easily, and yet be accurately centred on leaving the muzzle when fired. Several experimental guns were rifled on this system, but it was used only in the Royal Navy for a few 6.3 in guns. Meanwhile, a few ships were armed with 9 in/100 pounder smooth-bores and with 10.5 in/150 pounders. Of these about only fifty and a dozen, respectively, were made. Both were built up with wrought-iron coils on the Armstrong system. It was orig-

inally intended to use round shot at high velocity for close range attack on armour, the guns being rifled on the shunt system for longer ranges. However, the longitudinal strength of the gun was inadequate for this, and all those in service remained as smooth-bores. With steel shot at 200 yards, the 100 pounder was just capable of piercing $5\frac{1}{2}$ in, while the 150 pounder was a little better (Table 2). Armstrong's company also built a much larger gun during this period — the famous 'Big Will'. This was a built-up gun, 22.9 tons in weight, with a bore of 13.3 in diameter and 10.9 calibres long, rifled with 10 shunt grooves of a uniform twist of 1 in 65. With the normal charge of 70 lb and spherical steel shot of 344 lb, the muzzle velocity was about 1550 ft/sec. With an elongated projectile of 570 lb it gave 1200 ft/sec. Charges of 90–100 lb were occasionally fired. In December 1863, the gun sent a 612 lb steel shell with a cast-iron head, fired with a 70 lb charge, through the *Warrior* target at 970 yards, making a hole of about 4 square feet. The problem of mounting such a gun afloat had not yet been solved in Britain, and none of the guns already mentioned really satisfied the demands of broadside ironclad armament. In performance and weight the 7 in Whitworth was the nearest.

Between 1863 and 1865 two long series of trials were held. The first trial was between the Armstrong wedge BL, the shunt RML, and the all-steel Whitworth RML, using only 12 pounder and 70 pounder guns. The polygroove rifling and lead-coated projectile were disliked, as was the all-steel gun. Wrought iron, unlike steel, gave some warning of a burst. The apparent simplicity of a ML was preferred and the shunt RML gun considered the best of the three guns tried. The other trials were between 7 in RMLs. All had steel barrels and wrought-iron coils on the Armstrong system, but different rifling. These were: the French groove with increasing twist; the shunt groove; the Lancaster oval bore; Scott's method with a rifled projectile; and

Jeffery's and Britten's, both using lead at the base of the projectile. All except the French groove were uniform twist. The French system was preferred, and, slightly modified, became the standard Woolwich groove. In 1865, 9 in and 7 in RML guns constructed on the Armstrong system, with Woolwich rifling, were introduced and an 8 in gun in the following year. The 8 in and 9 in guns had increasing twist and the production 7 in uniform twist (Table 2). These guns were stronger than the French guns of 1864 and could fire charges of a more powerful powder to give a higher muzzle velocity. However, the same problems of barrel scoring and high localised loads on the rifling studs occurred, and, although the muzzle-loading system had an apparent simplicity, which seemed advantageous at the time, its adoption was to prove most unfortunate.

During this period the United States, for the most part, still used cast-iron smooth-bores. The 11 in Dahlgren shell gun could stand a 30 lb charge, giving a muzzle velocity of about 1330 ft/sec with a 186 lb wrought-iron shot. The standard gun in most of the monitors was the 15 in Dahlgren, with the 15 in Rodman gun used for coast defence. Both were cast hollow, with a water-cooled core. There were various models, weighing between 19–22 tons, the first gun being finished in 1860. In trials in Britain, a Rodman gun stood a 100 lb charge, which gave a muzzle velocity of 1545 ft/sec to a 452 lb cast-iron shot. At 70 yards a very good quality cast-iron shot pierced an 8 in wrought-iron plate with 18 in backing when fired with this charge. In later years a charge of this weight was allowed in service for 20 rounds using a slow-burning powder, but in this period 60 lb was the service maximum. With this latter charge, cast-iron shot would pierce the *Warrior* at 100 yards, but not the *Lord Warden* or *Bellerophon* at any range. Hardened steel shot (498 lb), however, would do so at very short ranges. These were the largest guns, actually mounted afloat during this period by a considerable margin. Although not as effective as a rifled

gun of the same weight, they were a fine culmination to the long history of cast-iron smooth-bores. Details of a model mounted in the later Monitors are given in table 2. 20 inch cast-iron smooth-bores were to be mounted in the monitor *Puritan* which was, however, never completed. A few of these guns were made in 1864. One, weighing 52 tons and 10.5 calibres long, later fired a 1072 lb shot with a 200 lb charge, giving a calculated muzzle velocity of 1370 ft/sec. Two of these 20 in pieces were later sold to Peru and mounted in the coastal batteries at Callao, where they caused the Chileans some apprehension.

The power of heavy guns increased greatly between 1866 and 1876. Improved techniques in construction and better quality materials, in particular more reliable steel, were partly responsible for this. The other main factor was the gradual introduction of a larger grain size in the black powder propellants. This led to a less rapid pressure rise, and a considerable reduction in peak pressure for the same weight of charge. Thus, for the same pressure maximum, a larger charge could be used, resulting in higher muzzle velocities. The best naval gun of 1866, the 9 in RML with a 43 lb charge of RLG (Rifled Large Grain) powder of about $\frac{1}{8}$ in linear grain size, fired a 250 lb projectile with a muzzle velocity of 1420 ft/sec. If a copper gas-check was fixed to the shot to prevent escape of gas, the projectile, now weighing 256 lb, would achieve a muzzle velocity of 1440 ft/sec. With 75 lb charges of P^2 (pebble) powder ($1\frac{1}{2}$ in cubes), over 1600 ft/sec could be obtained using gas-checks. These larger charges would, however, have led to troubles in magazine space and the need to strengthen gun-mountings, so that for this particular gun the largest service charge was 50 lb pebble. The United States were the first to adopt large grain powders in 1864, as a result of Rodman's work, reported in 1861. Their 'Mamouth' powder was irregular in shape and about 0.8 in diameter. Such powders allowed the use of the very heavy charges, latterly fired in the unhooped cast-iron 15 in and 20 in smooth–

bores. Russia, as a result of a mission to the United States during the American Civil War, was the next country to introduce hexagonal prismatic powder, followed by Germany. In Britain, prismatic powder was introduced in 1870–71, and in France, the somewhat similar Belgian Watteren powder was introduced at the same time. To obtain optimum results these powders required longer guns, chambers of larger diameter than the bore, and a means of preventing gas flow past the projectile, all of which went against the Woolwich muzzle loader. The continued use of wrought iron as opposed to all steel construction, was a further point against the British gun. Whereas, in 1866, it had been clearly the best type of naval gun, by 1876 the steel Krupp breech-loader had taken the lead. In fairness to the Woolwich RML it must be said that, when considering guns actually in completed ships, the superiority of the steel BL was only obvious at the end of this period.

The first Woolwich 7 in, 8 in, and 9 in guns were built on the Armstrong system, with a wrought iron or steel tube, a forged wrought iron breech-piece over the end of the tube and a series of shrunk-on welded wrought iron coils. This construction was retained by Armstrong, with improved steel tubes, through the forged breech-piece was eliminated. From 1866–7 a cheaper construction was introduced at Woolwich by R S Frazer, Manager of the Royal Gun Factory. In the prototype, the numerous wrought-iron coils were replaced by a single, thicker layer, the breech-piece being retained. Only a few guns were made in this way. A little later, the breech-piece was eliminated and a single heavy wrought iron coil shrunk over the breech, with the usual wrought iron tube (B tube) over the muzzle end of the inner steel A tube. There were, however, serious doubts as to the quality of welds in heavy wrought iron coils, so that, from 1869, the single breech coil was replaced by an inner and outer coil and this type of construction was used at Woolwich for the remainder of the

period. Data for these guns is given in table 3.

One experimental 13.3 in gun has already been described. Four more 23 ton 13 in weapons, to be rifled on the shunt system, were ordered in 1864. By 1867 only two survived, the others having failed due to split 'A' tubes, probably owing to the sharp corners of the shunt rifling combined with a tube which had not been heat treated. These two guns were re-rifled on the Woolwich system with uniform twist and a bore of 13.05 in. By this time a 12 in gun had been adopted and the two 13 in guns were not used in service. The 12 in gun, always known as 25 ton (although the first made weighed only $23\frac{1}{2}$ ton), was recommended in 1864 as likely to have a higher muzzle velocity than the 13 in. The first guns were completed in 1866. The rifling had nine Woolwich grooves with increasing twist from 1 in 100 to 1 in 50 (1 revolution in 50 calibres), but was found to be inadequate, and the 12 in/ 25 ton was less accurate than other Woolwich RMLs in consequence. This gun was first used at sea in the *Monarch* in June 1869. Meanwhile, a 10 in 18 ton gun was proposed in 1865, and first mounted in the *Hercules*, completed in November 1868. This gun, with seven rifling grooves with a twist of 1 in 100 increasing to 1 in 40, was widely used by the Navy and for coastal defence. An 11 in/25 ton gun was proposed in 1867 as an alternative to the 12 in, but, in October 1870, it was decided to complete the first guns of this calibre at 11 in instead of 12 in. The rifling had nine grooves with a twist of 0 to 1 in 35. In the Navy the gun was only used in the *Alexandra* and *Temeraire*, completed in 1877, although it saw service in coastal defence.

The next new gun, the 12 in/35 ton, was primarily designed for turret ships, the first being completed in February 1871. Rifled as the 11 in, it was rather too short and, though mounted in the *Devastation* and *Thunderer*, only 15 guns were made. The longer 38 ton gun that followed was tried with 12 in and 12.5 in bores and approved with the latter bore in November 1874. The first two guns afloat, in the fore-turret of *Thunderer*, were however left at 12 in to use the same ammunition as the 12 in/35 ton in the after-turret. After the double-loading and bursting of one of her 38 ton guns in 1879, the *Thunderer* was fitted with two 12.5 in guns. This type was also mounted in *Dreadnought*, completed in 1879. Rifling was the same as in the 11 in and, like the other RML guns so far described, the 12.5 in Mark I was unchambered. (It may be noted that the Armstrong 12.5 in/38 ton was first mounted in two Chinese gunboats, completed in 1877.) The Mark II 12.5 in mounted in the *Ajax* class and the rearmed *Dreadnought*, had a 14 in diameter chamber, but this was not adopted until 1880.

In 1872–3 both Armstrong's company and the Royal Gun Factory, at Woolwich, had prepared designs for a 60 ton 15 in gun. Both developed larger guns however — Woolwich a 16 in/80 ton gun and Armstrong a 17.7 in/100 ton. The first experimental 80 ton gun was made in March 1874 and completed in September 1875. Originally tested as unchambered 14.5 in, it was successively bored to 15 in, chambered to 16 in diameter, bored to 16 in and finally chambered to 18 in. The construction of this gun was based on the usual Woolwich principles with a total of five coils, double over the breech where the outside diameter was 72 inches. The two additional coils were shrunk-on between the 'B' tube and the breech coils, on account of the increased length. The rifling was on the Woolwich system with 11 grooves, 0 to 1 in 35 twist. To prevent erosion from gas leaking past the projectile, a copper gas-check was fitted to the latter, which expanded against the bore by gas pressure on firing. This device was not formally introduced into British service until August 1878, remaining 'experimental' until then. In December 1876, when the gun was an unchambered 16 in, a 30 in long crack was discovered in a rifling groove about 6 feet from the bottom of the bore and an incipient crack in an opposite groove. This

indicated that a different system of rifling was needed in the four guns being built for the *Inflexible*. In any case the Woolwich system was not conducive to accuracy of fire, as the projectile was not accurately centred in the bore. Tests were carried out on one gun, bored to 15.5 in, with polygroove rifling, rotation of the projectile was by a copper gas-check as in the 17.7 in Armstrong. It was eventually decided, in August 1878, after serious delays due to the development of very high pressures, probably from irregular ignition of the charge, to complete the four guns as 16 in bore with a chamber 59.6 in long, 18 in diameter, with polygroove rifling. The 33 grooves were 1.0 in × 0.1 in deep with increasing twist of 0 to 1 in 50 calibres. Rotation was by a copper gas-check. The delivery of the guns was delayed and *Inflexible* was not completed until October 1881 — the only ship to mount 16 in RMLs.

The Italians ordered Armstrong 100 ton guns in July 1874 and the first gun was completed as an unchambered 17 in and sent to Italy in July 1876. After successful trials at Spezia in late 1876, it was returned to England for boring, to 17.7 in, and chambering, and again tested in Italy in the spring of 1878. The other seven guns required for the *Duilio* and *Dandolo* were begun in June 1876. Apparently, as a result of the satisfactory tests of the Woolwich 80 ton, the Italian authorities decided to take the risk of further construction, prior to the first trials of the 100 ton gun. The four guns intended for the *Dandolo* were bought in 1878 for the defence of Malta and Gibraltar and a further four were sent to Italy. The steel inner tube was in two parts, joined by a shrunk ring, and a total of 17 wrought iron coils, including the trunnion-ring, were shrunk on, there being three layers over the breech with a maximum diameter of $76\frac{1}{2}$ inches. The chamber was 59.72 in long, 19.7 in diameter and the polygroove rifling, with a twist of 1 in 150 to 1 in 50, had 28 grooves, 1.1 in × 0.125 in deep. The projectile was rotated by a copper gas-check. In March

1880 the tube of one of the *Duilio* guns fractured at the base of the cone leading to the chamber, and the outer coils pulled apart just behind the trunnions. This happened with a 551 lb charge of the very large grain Italian Fossano powder, and the accident was attributed to irregular ignition of the charge, leading to abnormal pressures. In future full charges were reduced to 507 lbs. The longitudinal strength of the gun was considered fully adequate, but, whether because of this particular failure or for other reasons, the gun was considered longitudinally weak in the British service and limited to the same 450 lb charge as the 16 in. The 100 ton Armstrong gun was the most powerful muzzle-loader ever made, though Woolwich had a preliminary design for a gun of about 20 in bore. However, it was not likely that such a gun would be constructed.

The main development of heavy guns in this period was the rise of the Krupps' steel empire in Essen, Germany. As Germany was not at that time a major naval power, the improvement in guns was not so closely linked to ships, as in Britain, and armament of the warships of other countries must be included. Some of the first Krupp guns were made from a single steel forging, but all those considered here were strengthened by the addition of forged steel hoops. In the first of these guns, the steel tube, which had the breech slot cut into it, was reinforced by one to three layers of hoops over the breech-end. In the later construction, this tube ended at the front of the breech slot, which was then cut in a steel jacket shrunk over the rear part of the tube. The layers of hoops were shrunk over the jacket. In some guns, particularly those built to Krupp plans at the Russian Obuchov works, the hoops extended past the jacket to the muzzle. The breech was closed by a horizontal wedge working in the slot, the front face being flat and perpendicular to the axis of the bore, while the rear face was semicylindrical and inclined at about 2° to the front face. This type of wedge was in general use in heavy Krupp guns, though other types were used in some of the lighter and older guns. Obturation was by a steel 'Broadwell' ring pressed, by the powder gases on firing, against a recess in the chamber and against a steel face plate on the wedge. In most of the guns the chamber was slightly greater in diameter than the bore. The number of rifling grooves ranged from 32 to 40 in the older heavy guns and 48 to 90 in those made at the end of the period. The older type of groove had a depth of about $\frac{1}{8}$ in and a width of approximately $\frac{3}{4}$ in decreasing towards the muzzle. This was used with lead-coated projectiles, and, especially in those for some early guns, the coating might weigh 20% of the projectile weight, though later this was much reduced. The later type of groove was typically about 0.075 in deep, with a constant width of about $\frac{3}{8}$ in and the projectiles had copper driving bands. The rifling was generally of uniform twist, 1 rotation in 60–70 calibres with the earlier guns, and later 1 in 45. Krupp described the length of the guns by the total length in calibres, but for uniformity with other types, in this paper the length of the bore from the front face of the breech block is taken, which is about three calibres less than the Krupp figure.

At the Paris Exhibition in 1867, Krupp showed an all steel 14 in gun about 12 calibres in bore length, and weighing 50 tons. This had not been fired at the time, and was only a low velocity gun. In 1868, a series of tests were made on a 9.3 in Krupp gun against the Armstrong 9 in RML. Initially the Armstrong gun was far superior due to the low muzzle velocity, about 1150 ft/sec, of the Krupp 9.3 in. This disadvantage, which arose from the less powerful German powder and the increased projectile weight due to the heavy lead coating, was corrected by a heavier charge of prismatic powder and a thinner lead coat. Subsequently, the Krupp gun was judged to be superior. Though latterly very reliable, a number of failures occurred in early Krupp guns, probably due to the

Table 3

Gun	Length of bore *calibre*	Weight of gun *ton*	Nominal weight of projectile *lb*	Charge *lb*		Muzzle velocity *ft/sec*
10 in RML	14.5	18	400	60	RLG	1298
				70	pebble	1364
11 in RML	13.2	25	536	85	pebble	1315
12 in/25 ton RML	12	23½–25	600	67	pebble	1180
				85	pebble	1297
12 in/35 ton RML	13.5	35	698	110	pebble	1300
12 in/38 ton	16.5	38	698	110	pebble	1430
			original guns in 'Thunderer' only			
12.5 in/38 ton RML	15.8	38	818	130	P²	1415 gas-check
				165	prism	1442
16 in/80 ton	18	81	1700	450	prism	1604
17.7 in/100 ton Armstrong RML	20.5	102	2000	450	prism	1548
			British land-service charge			
				507	fossano	1641
			Italian service charge			
				551	fossano	1700
12 in Krupp	18.9	35	717	159	prism	1601
14 in Krupp	21.8	51.3	1157	253	prism	1650
15.75 in Krupp	21.8	71	1712	452	prism	1648

uncertain quality of steel forgings at the time. In about 1870, Krupp guns were made in bore lengths of approximately 15, 17 and 19 calibres.

Of the later guns, with copper-banded projectiles, a 12 in/19 cal gun was first mounted afloat in the German armoured gun-boats of the *Wespe* class, completed toward the end of 1876 and also later used in the Danish *Helgoland*. On a muzzle energy per unit weight basis, these guns were generally better than unchambered Woolwich guns with gas-checks. In 1875–6 a 14 in/51 ton gun, and a 15.7 in/71 ton gun, both 22 calibres in bore length, were built by Krupp and were demonstrated very effectively at Meppen in 1878 and 1879 respectively. Unlike other Krupp guns already mentioned, the 15.7 in gun had a chamber considerably larger in diameter than the bore, namely 17.32 in by 61.37 in long. The 14 in had, however, a small chamber, only 42.67 in by 14.37 in. These guns had an appreciably higher muzzle-energy per ton weight than either the Woolwich 16 in or the Armstrong 17.7 in. The 15.7 in was never

mounted afloat, but the small Danish coastal-defence ship *Trodenskjold*, completed in 1883, carried a single 14 in of the same weight and length, though with a larger chamber.

Table 3 gives data for most of the British and Krupp guns already discussed. In several guns, other types of charge, often giving a different performance, were also used. Between 1870–1890 there was a marked development in gunpowder propellant charges.

In Britain a few guns had been mounted in many ships for saluting and other purposes. There were usually the 3.7 in BL, but some had either the 4.7 in BL or the 6.3 in RML. None of these were powerful guns and were not mounted in secondary batteries till the *Ajax* class, completed in 1883, was fitted with two 6 in BL of a new type. In 1880 the Captain of the *Excellent* wanted six 6 in BL in the *Inflexible*, but his request was turned down and her secondary armament was only six 3.7 in guns.

The period from 1877–1892 saw marked progress in heavy guns. Although weapons of

16–17 in bore were mounted afloat, by 1890 the trend was to smaller diameters of 12–13.5 in, with increased length and muzzle velocity. In 1877 the largest guns were only 20–22 calibres long giving muzzle velocities of about 1650 ft/sec. By 1892 lengths of 30–35 calibres and velocities of over 2000 ft/sec were usual and the most advanced guns, notably those made in France, were 45 calibres in length with velocities of over 2500 ft/sec at the muzzle. Once again improvements in propellants were responsible for the increased performance. By about 1881 black powder was used everywhere in prismatic form, the individual grains being generally about 1 in in height and $1\frac{3}{8}$ in over the sides, but sometimes 1 in more in height and width.

In 1882 'brown' or 'cocoa' powder was introduced in Germany. The composition of this was about 79% potassium nitrate, 18% charcoal and 3% sulphur (black powder was typically 75, 15, 10). The charcoal was made from partially carbonised rye-straw. The resultant powder in prism form was considerably slower burning than black powder and was adopted by all countries for heavy guns. Its disadvantage was the great weight of the charge. In the 16.25 in gun, this reached 960 lbs of SBC ('slow-burning cocoa', an improved variant with carefully controlled moisture content) in eight 120 lb sections, the heaviest propellant charge ever used in a ship-mounted gun. A more important advance was the first successful conversion of nitrocellulose into a propellant. Attempts to use it for charges had been made since Schonbein's discovery of nitrocellulose in 1845, but it was not until 1886 that success was achieved. Credit is usually given to Vieille in France, though many German writers credit Duttenhofer. Nitrocellulose/nitroglycerine compositions soon followed. The advantages of nitrocellulose over gunpowder were great. Smoke was very much less and the charge weight only a half to a third for the equivalent performance. Once the techniques had been mastered, the propellant was more stable than powder and

far less sensitive to shock and friction. The first heavy gun designed to use such 'smokeless' nitrocellulose powder was the French 1887 model, completed from 1891–2 onwards. In these the nitrocellulose was used in strip form.

Much difficulty and delay was experienced in building many of the heavy guns now used. The increased length required long hollow steel forgings of high quality. These were difficult to make and heat treat and also difficult to assemble by shrinkage. Many different designs were tried, the tendency being to fewer and longer forgings as the period advanced. The disadvantages of the muzzle-loader increased sharply with greater length, and Britain adopted breech-loaders with the French type of interrupted screw breech.

The rifled muzzle-loader reached its peak with the Armstrong 17.7 in, of 20.5 calibres bore length, though longer muzzle-loaders were made in the late 1870s. In 1878, an Armstrong 11 in of 23 calibres appeared and this type was mounted in eight Chinese gunboats (Table 4). This gun was never adopted in the British service, but two experimental 10.4 in/26 calibre guns were made in 1880 for trials on land, with a breech loader of similar dimensions, which proved superior. In 1880 the chambered Mark II version of the 12.5 in RML was introduced. Apart from the chambers, these guns were the same as the Mark I, and many were, in fact, conversions. The gun's performance was, however, much improved and the final figures obtained will be found in Table 4. The favour always shown towards muzzle-loaders was adversely affected by an accident to a 12 in/38 ton gun in the *Thunderer* on 2 January 1879. Both the fore-turret guns should have been fired together, but it was not noticed that one had failed to fire. Both guns were reloaded and on firing the second time, the gun burst, killing eleven men and wounding thirty. The gun had both the original 110 lb pebble charge and 688 lb palliser projectile, with the second charge of 85 lb pebble charge and 575 lb common shell. The sister gun subsequently burst in a very similar manner when experimentally loaded in this way. It may seem incredible that the different behaviour in recoil, or the different rammer travel were not noticed, but contemporary opinion considered this perfectly possible with hydraulically-worked guns. It was impossible for such an accident to occur with breech-loading guns.

In June 1879 the Committee on Ordnance recommended that a 12 in/43 ton BL be constructed with a steel barrel and wrought-iron coils. Elswick 8 in and 6 in BLs of this type of construction had so far performed very satisfactory on trials, and by early 1881 the 12 in gun had passed proof. The proposed charge of 285 lb P^2 (large pebble) was expected to give a muzzle velocity of 2001 ft/sec with a 714 lb projectile, but this charge was too bulky to enter the chamber. With 285 lb of prismatic powder the velocity was only 1840 ft/sec. It was hoped, however, that if the breech opening was enlarged to the full chamber diameter, a 400 lb slow burning charge might be used to give about 2200 ft/sec. In 1881 a new design was prepared at Woolwich and in 1882 it was decided to make the first batch of 12 in/43 ton BL guns. The Mark I was trunnioned for land service and the Mark II trunnionless for the Navy, the two being otherwise identical. The detailed construction of these guns was unfortunate. It consisted of a steel 'A' tube, the end of which was threaded for the breechblock. From the fore end of the chamber, to 8 in from the muzzle, three welded steel coils were shrunk on. A wrought-iron coiled breech piece was shrunk over the chamber and breech-end of the gun with a wrought-iron coil in front, over part of the steel coil layer. Finally, a long wrought-iron coil was shrunk over the breech-piece. The Armstrong Company objected to the design of these guns, which they were about to make, in May 1882. It is obvious, nowadays, that welded steel coils were a poor substitute for forged hoops, and wrought iron inferior, as a material, to steel. It may be noted too, that it was not until 1884 that the

Ordnance Committee recommended that all tubes and solid forgings were to be annealed prior to hardening and adequately tempered after quenching. The absence of such treatment was the cause of failure in pre-service firing and the charge was limited to 295 lb of brown powder. Even so, on 4 May 1886, the 'A' tube of one of these guns, mounted in *Collingwood*, broke in pieces in front of the last hoop when fired with a three-quarter charge. The whole batch was withdrawn for chase-hooping and removed from the Navy, being converted into trunnioned guns for land service. The guns were replaced by those made entirely from forged steel. Of these there were four different patterns in the Navy — Marks III, IV, V and Vw.

Mark III was an Elswick 1882 design, originally trunnioned for land service, and altered to a trunnionless gun, chase-hooped to the muzzle. There was an 'A' tube, a breechpiece into which the block screwed, and a total of 27 hoops with three layers over the breechpiece and one layer at the muzzle. Use was made of yellow metal which ran into grooves to increase the longitudinal strength of the assembly. Mark IV was also altered by chase-hooping to the muzzle. This was a Woolwich 1884 design with an 'A' tube and breechpiece, which took the breechblock, and a total of 14 heavier hoops. There were two layers over the breechpiece and projections were used to lock the hoops together. Mark V was a Woolwich 1885 design. This was not chase-hooped to the muzzle, but the inside of the 'A' tube had a liner at the breech-end and an 'alpha' tube at the muzzle. Over the 'A' tube was the breechpiece taking the block, and a 'B' tube, which did not extend to the muzzle. The remaining two layers were composed of two hoops. Mark Vw was a Whitworth variant, chase-hooped to the muzzle, but without liner or alpha tube. Data for these were similar and will be found in Table 4.

The 13.5 in guns, which were all trunnionless, were built up from steel forgings only. There were four marks and seven submarks. The weight was about 67 tons except for the

Mark I guns, of which only four were made for the *Rodney*, which weighed $68\frac{3}{4}$ tons. Most of the guns were Mark IIIs and its various submarks. Mark I was an 1884 design having an 'A' tube with, originally, a half liner and later both bore and chamber liners. Over the 'A' tube was a breechpiece, into which the block screwed, a 'B' tube and four hoops to the muzzle. Two layers, totalling eight hoops, were shrunk over the breech-end. Mark II was an 1885 design built up in a similar way to the 12 in. Mark V and Mark IIa only differed in having separate bore and chamber liners. Some liners cracked in proof, and the later marks were only lined when worn. Mark III was an 1888 design, with an 'A' tube, two 'B' tubes extending from the front end of the chamber to the muzzle, a breechpiece taking the block which extended partially over the first 'B' tube, and a 'C' tube in front of it. Over the breech-end were two layers, comprising two hoops and a long hoop or 'jacket'. The 'A' tube had shoulders nearly to the muzzle, which enabled better provision for longitudinal strength. Submarks a to e were introduced to use parts already made for Mark II guns and generally only differed in having some of the longer hoops replaced by two short ones. Mark IIIf, which was an 1889 design, had the breechpieces shrunk on from the muzzle-end and secured at the breech-end by a screwed steel bush. The one 'B' tube was replaced by two hoops and the 'A' tube had shoulders over the powder chamber to give greater longitudinal strength. Mark IV dated from 1890–1. The parts were built-up from the muzzle-end and secured in position by screwed steel bushes. They comprised an 'A' tube, a bush, taking the breech-block which screwed into the breechpiece and into the '1/ C' hoop, a breech-piece, 'B' hoop and 'B' tube to the muzzle. Next came the '1C' hoop, 'C' tube and '2C' hoop, and, at the breech-end, were two more layers as in the Mark III. This was the final type of Woolwich built up gun. As with the 12 in the various marks had similar data which are given in Table 4.

The huge 16.25 in gun was an Armstrong, designed and built at Elswick. It was developed from the 17 in which had been made in five different versions for Italy. The most important difference from the last of the 17 in guns was that the 'A' tube was in one piece and not in two. In all, twelve 16.25 in guns were made and variations were such that the guns were known by their serial numbers and not by different marks. The first two guns issued for *Benbow* had an 'A' tube, with a breechpiece taking the block and 17 hoops to the muzzle, over it. Next came a layer of eleven hoops, then one of eight and finally one of seven. The breechpiece was screwed to the 'A' tube longitudinally by a serrated metal ring, and the outer hoops were secured by shoulders, with some use of yellow metal also. The rifling grooves in these two guns were 0.04 in deep which, with full charges, tended to strip, so that in subsequent guns the depth was increased to 0.06 in. After proof and firing trials some of the next guns, as originally made, drooped and bent slightly, and some of the forward hoops moved apart a little.

The next seven guns were therefore modified by the use of fewer, longer hoops. In detail there were five different assemblies, but it can be said that four guns had a long hoop at the end of the second layer of hoops. The other three guns were more drastically altered. The 17 hoops of the first layer were replaced by three hoops and three tubes, and the number of hoops in the second layer was also reduced. In the last three guns made, the 'A' tube was supported by the breechpiece, which took the block, and by three tubes extending to the muzzle. Next came two layers each consisting of three long hoops and a final layer of two long hoops and a locking ring. This new design gave improved longitudinal and girder strength and also increased the circumferential strength (Table 4). The full charge life of the 16.25 in is usually given as 75 rounds before re-lining, as compared with 105 for the 13.5 in.

British guns of this period had large diameter chambers, the sizes being:

16.25 in:	$84\frac{1}{2}$ in $\times$ $21\frac{1}{8}$ in;
13.5 in:	$66\frac{1}{2}$ in $\times$ 18 in;
12 in Mk II:	55.8 in $\times$ $14\frac{3}{4}$ in;
12 in Mk III–V:	48 in $\times$ 16 in.

The rifling was of polygroove type with grooves of various forms, from 0.05–0.06 in deep, except in the first 16.25 in gun. The number of grooves was equal to 4 times the calibre in inches except for the 16.25 in which had 78. Increasing twist was used, the final helix being 1 in 30 (in the 12 in, 1 in 35), though there were differences in the initial twist and in the position where the twist stopped increasing before the muzzle, where the rifling was usually of constant helix. Copper driving bands were used, as was now standard practice in all BL guns. The breech blocks had four interruptions in the 12 in, five in 13.5 in and six in the 16.25 in and were all worked hydraulically. De Bange obturators, in which a mushroom-headed piston was driven back against an asbestos-tallow gasket on firing, were standard from 1882. During this period successful wire-wound guns were made at Elswick and Woolwich, but these are more conveniently described in the next period, when wire winding became normal practice for British heavy guns. It must be noted that during the years 1887–1891 the Navy assumed responsibility for its own guns — the Naval Ordnance Department being set up in 1891.

The many varied designs described show the difficulties involved at this time in building relatively long steel guns. In France similar troubles were experienced and a great advance was made with the 1887 model. For the first time these were designed to use nitrocellulose propellants instead of brown powder. 13.4 in, 12 in and 10.8 in guns were made, of 44–45 calibres length. The 13.4 in was the exception and was limited to 41.5 calibres as the naval factory at Ruelle had not the machinery to complete a longer gun. The construction was simple, namely a thick 'A' tube, into which the breech block screwed, and a breechpiece over it, with chase-hoops beyond it to the muzzle, the front hoops being screwed to the 'A' tube. Over the breechpiece was one row of hoops. The first 13.4 in, which only weighed 57 tons, bent on trials and the gun had to be reconstructed to a modified design, with a longer breechpiece and a total weight of 60.6 tons. Some of the 12 in and 10.8 in guns were also modified. Whatever the difficulties with these guns, they mark the division between the older and newer type of heavy gun. Details of the 1887 guns are given in Table 4 where the high muzzle velocity and large energy per unit weight, contrast with those of other guns.

The chamber in the 12 in was 68.3 in $\times$ 13.94 in, small by British standards for a gun of its power. The number of rifling grooves was 3 times the calibre in centimetres, and the depth of the groove in heavy guns was from 0.067 in–0.079 in. Uniform twist of 1 in 36 was employed. Breechblocks had four interruptions and had the Manz hand-operated breech mechanism where one man, by continuous rotation of a wheel, could open the breech in 10 seconds. Although the French army and the Canet factory (for the guns sold abroad) favoured De Bange obturators, the French navy used Broadwell rings during this period.

The only new heavy Krupp guns in the German navy were the 11 in/37 and 32 calibre guns in the *Brandenburg* class. The Russians introduced 12 in/32 calibre guns made by Krupp and at Obuchov. A 12 in/32 calibre gun, 1880 model, was introduced in the Austrian navy. Much larger Krupp guns were made for coastal defence and the heaviest guns in the Dardanelles forts in 1915 were five 14 in/32 calibre guns weighing 80 tons. These were the 1880 model ordered in 1885. Larger still were the four 15.75 in/32 calibre 119 ton guns built for Italy in 1885-6 and mounted in twin turrets at Spezia and Taranto. These guns were also of the 1880 model. The Taranto guns were for a time mounted afloat, one each in the small Italian gunboats *Castore*

Table 4

Gun	Length of bore _calibre_	Weight of gun _ton_	Nominal weight of projectile _lb_	Charge _lb_		Muzzle velocity _ft/sec_
British guns						
12.5 in MK II RML	15.8	38	818	200	EXE	1575
11 in Armstrong RML	23.2	35	535	235	pebble	1875
12 in MK III–Vw	25.2	45/46	714	295	pebble	1914
		(12 in MK II originally 25.1 cals, 43 tons, other details same)				
13.5 in	30	67/69	1250	630	SBC	2016
16.25 in	30	111	1800	960		2087
French guns						
10.8 in (1887)	45	34.4/37	562	149	NC	2559
12 in (1887)	44.3	45/48.5	750	196		2559
13.4 in (1887)	41.5	60.6	1080	243		2428
Krupp guns						
12 in Russian 32 cal	31.9	55.2	729	347		2090
12 in Austrian	32	47.2	1003	309		1755
11 in German KL/35	32	43.2	529	317		2247
11 in German KL/40	37	43.3	529	317		2346
15.75 in Italian	31.75	119	2028	727		1804
Above are specified trial minima for 15.75 in.						
Maximum figures are quoted as:			2315	847		1900

Propellants were brown prismatic powder unless otherwise noted. NC is nitrocellulose, SBC, slow-burning cocoa and EXE a prismatic powder of approximate composition 2/3 brown; 1/3 black.

and *Polluce* of 530 tons launched at Pozzuoli in 1889. Krupp lists, at the end of the period, show 15.75 in/37 calibre guns and 11 in and 12 in guns of 47 calibres. These last guns were much heavier than corresponding French 1887 guns and are listed with about the same performance.

The same general construction was used in all guns. The 'A' tube extended from the front face of the breech-slot, which was cut in a heavy breechpiece shrunk over the 'A' tube. Chase hoops extended to the muzzle from the forward end of the breechpiece, and over this last were one or more layers of hoops. In the 15.75 in guns built for Italy, there were three such layers and in the 12 in/32 calibre Obuchov, four layers, of which one extended well over the chase hoops towards the muzzle. The Russian 12 in guns were particularly heavy compared to guns of other navies. In all Krupp heavy guns the breechblock was the single 'cylinder-prismatic' wedge with obtu-

ration by Broadwell ring. Chambers were usually long, extending to about 6 calibres or more in some guns and of not much greater diameter than the bore, though in the 119 ton gun the chamber diameter was 18 in. At the beginning of the period the rifling twist was uniform, but later an increasing twist was used, with a final twist of about 1 in 25 or 1 in 45 in Russian 12 in guns. The number of grooves was usually in the range of $2\frac{1}{4}$–3 times the calibre in centimetres, and the depth, 0.06 in–0.08 in (Table 4).

In smaller calibres the most notable event of the period was the introduction of the quick-firing (QF) gun. This evolved from the light anti-torpedo-boat guns introduced in the 1880s. Its adoption in larger sizes, as standard warship armament, was due to Armstrongs at Elswick. In 1887 a 4.7 in QF gun was submitted for test and fired 10 rounds in $47\frac{1}{2}$ seconds, almost 8 times faster than the 5 in BL. It was decided to mount the 4.7 in in the *Trafalgar* class in 1890 and the 6 in QF followed. The features of the QF gun were:

1. A metal cartridge case, which included a primer so that no tube was needed. The cartridge case also did away with both the need to ram the projectile separately and sponging.
2. A rapid breechblock action with mechanical extraction of the cartridge case.
3. Sights which did not recoil with the gun and recoiling and running-out arrangements which did not disturb the aim of the gun-layer.

In Elswick guns the breechblock was coned so that it could be swung back quickly, but in other patterns a cylindrical block was usual. Krupp retained the wedge breech for QF guns. In Britain both 4.7 in and 6 in guns had separate projectiles and cartridge cases, but elsewhere fixed ammunition was normal up to 4.7 in at this time, and some French 5.5 in and German 5.9 in guns used it.

As shown by the original 4.7 in above, QF guns could attain high rates of fire and even with a 3-motion breech mechanism the 6 in Elswick gun could fire 18 rounds in three minutes. Average rates would be rather better than $3\frac{1}{2}$ rounds per minute for the 6 in and nearly $5\frac{1}{2}$ rounds per minute for the 4.7 in. These rates of fire called for an increased ammunition allowance and whereas with BL secondary guns, 100 rounds per gun was an ample figure, with the QF, 200 or more rounds per gun might be carried.

The 4.7 in and 6 in guns in the British Navy were 40 calibres, weighing approximately 2 and $6\frac{1}{2}$ tons, and fired 45 lb and 100 lb shells. With cordite, muzzle velocities were respectively 2125 and 2230 ft/sec, but in this period they had rather smaller powder charges and velocities were only 1786 and 1882 ft/sec.

The menace of torpedo attack was much to the fore in this period and it was considered essential to stop an attacking torpedo-boat at a minimum 400 yards range by gunfire. The increase in the power of attack was matched by the adoption of more powerful anti-torpedo-boat guns, though generally these lagged behind and tended to be of too small a calibre for their task. If used in attack, torpedo gun-boats could only be met by the quick firers of the secondary battery. The most widely used guns intended for fire against torpedo-boats at first was the 37 mm Hotchkiss revolver, which was also intended to sweep the upper-deck of a battleship in close-action from bulwarks or tops. First tried at Gâvre in 1873, this weapon was adopted by the French Navy in 1876–7. There were five barrels, revolved by a crank, firing in turn. The bore was only 20 calibres long, and the 1 lb shell had a muzzle velocity of 1318 ft/sec. The rate of fire could reach 70 per minute, but for aimed fire it dropped to 14. The gun weighed 462 lb and was mounted without recoil gear in a 'Y' shaped piece which fitted into a socket on the bulwarks or elsewhere. This gun was not used in the British Navy, the 1 in Nordenfelt being adopted instead. In the type mounted in battleships, there were four barrels fixed in a horizontal plane, with a breech mechanism worked by a lever, and hoppers holding 10 rounds per barrel. The length was 35 calibres and the projectile was a solid steel shot of $7\frac{1}{4}$ ozs with a muzzle-velocity of 1450 ft/sec. The maximum rate of fire was as high as 216 per minute, but for aimed fire it was 40. The gun was a little lighter than the Hotchkiss, but it had a much heavier mounting with elevating and training gear but without recoil mechanism. The Hotchkiss had the advantage of an explosive shell, but in the early 1880s it was clear that neither of these guns were adequate.

In 1881 the matter of a more powerful gun was taken up in Britain, and eventually in 1886 the 1 in Nordefelt was superseded as the main anti-torpedo-boat gun by the 57 mm 6 pounder Hotchkiss and Nordenfelt guns. These were single-barrelled guns, with a vertically sliding blockbreech in the Hotchkiss and a falling block in the Nordenfelt. They were 40–42.7 calibres in length and fired a steel-pointed, base-fused 6 lb shell with a muzzle velocity of 1818 ft/sec. Fixed ammunition was used and, with a crew of three men, the rate of aimed fire was 12 rounds per minute. The original 6 pounder mountings had no recoil gear and needed a very strong structure below them in consequence. Accordingly for use in the tops and other places unsuited to the 6 pounder a Hotchkiss 47 mm 3 pounder was introduced. This was similar to the larger gun and fired a 3.3 lb shell at 1873 ft/sec. The mountings had recoil gear as did the later 6 pounder mountings.

Germany took the wisest course and adopted a Krupp 88 mm (3.4 in) QF. This was a 27-calibre gun and fired a 15 lb shell at 2020 ft/sec. Some mountings had a seat for the gun-layer, which moved with the gun, so that he could train, elevate and fire. The rate of fire was said to be 20 aimed rounds a minute.

From 1893 heavy guns were widely standardised at 12 in and the exceptions were of smaller, not of larger, bore unless, as in some United States ships, guns of an older pattern were still used. Following the French lead with their 1887 model guns, described above, smokeless propellants were adopted and muzzle velocities increased. There was much variation between the different navies in the propellants used. In Britain cordite Mark I was used. Its composition when originally introduced, was 58% nitroglycerine, 37% nitrocellulose, and 5% petroleum jelly. It had a very high energy per unit weight but also produced very hot gases and a most undesirable erosive action on the gun. From 1902 this was replaced by MD cordite containing 30% nitroglycerine, which was far more satisfactory and more than doubled the life of a 12 in gun compared with the earlier propellant. In both cases the propellant 'grains' were in cylindrical rod ('cord') form. This was a markedly 'degressive' shape, i.e. the rate of gas evolution decreased with time and was at a maximum initially. Except for Japan which obtained propellants from Britain at this time, other countries favoured less 'degressive' forms. Of those navies using nitroglycerine/nitrocellulose compositions, Germany, after prolonged experiments, adopted a somewhat similar material to MD cordite, but in tubular form, while Italy used Ballistite. This originally contained 50% nitroglycerine, in strip form, though this was later reduced. This propellant had no petroleum jelly or other inert material incorporated in it and 50/50 Ballistite was more erosive than cordite Mark I. The other principal navies used nitrocellulose. The French favoured strips, as did the Russians, though by 1905, the Russians were making some use of the tube. The United States did not introduce smokeless propellant until 1898 and adopted a relatively short 'grain' with seven perforations. Nitrocellulose propellants gave cooler gases than the nitroglycerine/nitrocellulose types, but the energy per unit weight was less and larger charges were needed. Nitrocellulose propellants were also originally considered less stable, more liable to irregular burning and to 'back-flame' or 'flash-back' (ignition of residual inflammable gases present in the gun after firing). Flash-back was a danger with all 'smokeless' propellants, and by the end of the period, arrangements to clear the bore by either an air/water or an air blast, prior to opening the breech, were generally fitted. For a given gun, charge weight and maximum presure, cord gave a higher muzzle velocity than other propellant shapes. However, it was usually possible to devise a heavier charge of tube or strip, which would give the same maximum pressure as the cord charge and a greater muzzle velocity, though regularity might be reduced. All 'smokeless' charges were difficult to ignite and special ignitors, usually made from black powder, had to be provided.

'Smokeless' propellant was first used in a naval action during the Brazilian Civil War in 1893, when the cruiser *Tamandare* had Ballistite charges for her 6 in QF guns. One very serious danger with smokeless propellants was their tendency to decompose, in which case spontaneous explosion could occur.

Though charge weights were much smaller than with brown powder, the density of loading was also much less. Otherwise pressures would have been far too high, and chamber sizes were thus not appreciably reduced, and, for a given bore diameter, were generally increased. As a result, in spite of the increased length of the newer guns, the expansion ratio (total volume of bore divided by volume of chamber) was often reduced. This was particularly marked in the later British guns where the 40 calibre, 12 in MK IX had larger chamber volumes than the 13.5 in/30 calibre and the expansion ratio was appreciably less in the 12 in MK IX as compared with the older 13.5 in guns.

Wire-winding was now used for the new 12 in guns. It was not adopted by other countries, who continued to build their heavy guns entirely from steel forgings. There were certain advantages in the use of wire. The tensile strength was much higher than that of forgings available at the time, and it was far easier to check the quality of the material. Also it was easier to wind wire at a given tension than to shrink long forgings to precise limits. However, the wire layers contributed little to longitudinal strength or resistance to muzzle droop. In addition, although the earlier British wire-wound heavy guns were much lighter for a given muzzle energy than the previous British built-up guns, they were no lighter, on average, than the contemporary French guns, though they had a higher safety factor.

As long ago as 1875–6 the French had tried wire-wound guns made to Schultz designs. The first were small, only 90 mm (3.5 in) bore, but later, guns as large as 9.4 in were made. In 1882 a 13.4 in 48 ton gun was sent for trial. This had a steel 'A' tube wound with round wire of 0.118 in diameter and 127 tons per sq in tensile strength. There was a steel breech-piece, a steel trunnion hoop and two cast-iron jackets. The trunnion hoop and breechpiece were connected by 12 steel bolts. At the first trial round six bolts were broken, and the gun was modified, the bolts being replaced by a

steel jacket. Further trials were held in 1887–9, but at the 33rd round the gun parted transversely, though the pressure of 19.6 ton/sq in was not excessive, and the system of construction was abandoned. Krupp's did not favour the use of wire, but a British 10.2 in/29 calibre 21 ton gun was purchased from Armstrong's for trials in 1882. This had a heavy 'A' tube, wound to the muzzle with flat wire of 60 tons per sq in tensile strength, though the wire was not in a continuous coil. Over it were steel hoops and wrought-iron coils. This gun behaved well, and was later relined to 10 in. In the 1880s about twenty 10 in guns were made by Armstrong's, using wire-winding for use in other countries. Meanwhile, in 1884 Woolwich had designed a 9.2 in/40 calibre wire-wound gun. The original design would have weighed only $19\frac{1}{2}$ tons, but chase-hooping raised it to 22 tons. The wire did not extend over the whole length of the gun and was covered by two long hoops extending to the trunnion ring. This gun went for trials in 1886 when, with a 270 lb brown powder charge and a 380 lb projectile, the muzzle velocity was 2360 ft/sec. In 1887, the then remarkable range of 21,800 yards was attained at 45° elevation. The success of this gun was followed by the design of an experimental 69 ton 13.5 in gun in 1890–91. In 1892 manufacture of the wire-wound 6 in QF begun. The wire used was flat in section, 0.06 in × 0.25 in and of 90 to 110 tons/sq inch tensile strength.

In 1893 the design of the first wire-wound gun to be used as the main armament of a battleship was approved. This was the 12 in Mark VIII of 35.4 calibres and 46 tons. This gun was to be mounted in the *Majestic* and *Canopus* classes, a total of 15 ships. The construction consisted of an 'A' tube with an inner 'A' tube driven cold into it, which could be replaced when worn with relative ease; wire layers over the whole length, of varying thickness, and a jacket and 'B' tube, which did not overlap and had a 'C' ring screwed over the join. The exterior of the 'C' ring was grooved for securing to the saddle. The breech block had six interruptions to the thread and was contained by a breech bush screwing into the 'A' tube, located against the end of the inner 'A' tube. The 'A' tube and jacket were joined by a collar, which was shrunk onto the 'A' tube and screwed into the jacket at the breech-end. The breech mechanism was designed for hand operation, though some mountings had hydraulic power. The chamber was 70 in by 16 in maximum diameter, with Mark I cordite and was lengthened to 76.952 in for MD. The rifling was straight for nearly 6 ft at the breech and then increased in twist to 1 in 30 at the muzzle. There were 48 grooves of modified plain section, 0.607 in × 0.08 in, though rather larger where straight. Later re-lined guns were rifled with a uniform twist of 1 in 30.

The 12 in Mark IX of 40 calibres and 50 tons, which followed, was a Vickers design which was approved in 1898. This was mounted in the *Formidable, London, Duncan,* and *King Edward* classes, which numbered 22 battleships. The construction was similar to the Mark VIII, except that the 'A' tube was thickened at the expense of the inner 'A' tube, and the 'B' tube did not extend as near to the breech and had the end of the longer jacket shrunk over it, instead of a 'C' hoop over the join. The securing grooves were on the jacket. The Welin stepped-screw breechblock was introduced in this gun. This had twelve segments of which three were plain and nine threaded, three at each of three different diameters. The mechanism was hand operated except in the *King Edward* classes, in which it could be either hydraulic or hand operated. The chamber was originally 87.4 in by 17.5 in maximum diameter, later lengthened to 92.767 in for MD cordite. There were forty-eight rifling grooves, 0.62 in × 0.1 in, and the rifling was straight for 4 in and then of increasing twist to 1 in 30. As in the Mark VIII this was later changed on re-lining to a uniform twist of 1 in 30. However, the chamber was too large, causing a small expansion ratio of the propellant gases. Thus stress concentration occurred at the forward locating

shoulders between the inner and outer 'A' tubes leading to cracked inner 'A' tubes and closing in of the bore. This also occurred in the MK VIII gun and the cure (taper boring of the 'A' tube and taper turning of the outside of the inner 'A') was at least a decade away.

French guns of this period varied, but had certain common features. Breechblocks were cylindrical with four interruptions, and were hand operated. De Bange type obturators, universal in British made guns, were still not in use and obturation was by copper rings. Rifling was of uniform twist, 1 in 36 or later 1 in 45, and the number of grooves was 3 times the calibre in centimetres.

The M1893 12 in and 10.8 in guns were of 44–45 calibres. The 'A' tube was reinforced at the breech-end by two layers of short hoops which ended in a locking hoop screwed onto the tube. Over this part a long jacket was shrunk, connected to the '1B' tube by a screwed locking hoop, while the '2B' hoop, reaching to the muzzle, was screwed onto the 'A' tube. The bush for the breechblock screwed into the jacket and was located against the 'A' tube and its reinforcing hoops. The guns were mounted in the *Bouvet* and the *Massena*. Chamber sizes were similar to those in the M1887.

Both the 12 in and 10.8 in of the 1893–6 models were of 40 calibres. The 12 in gun was mounted in the *Charlemagne* class, *Iena* and *Suffren*, and the 10.8 in only in the *Henri IV*, though it was used to re-arm the *Calman, Indomptable* and *Requin*. These guns were generally similar to the 1893 model, but had larger breechblocks and very long chambers — 108.4 in × 12.6 in in the 12 in for example.

Although the Krupp gun tables listed powerful 11 in and 12 in guns of 42 and 47 calibres, by the end of this period the only new heavy gun mounted afloat in the German Navy was an 11 in of 37 calibres. This was considerably more powerful than the previous 11 in/37 calibres in the *Brandenburg* class, but of the same normal Krupp construction with

Table 5

Gun	Length of bore calibre	Weight of gun ton	Nominal weight of projectile lb	Charge lb		Muzzle velocity ft/sec
British guns						
12 in MK VIII	35.4	46.1	850	174	cordite MK I	2417
				200	cordite MD	
12 in MK IX	40	50.8	850	211	cordite MK I	2567
				246	cordite MD★	2561
(In King Edward class only)				254	cordite MD	2612
French guns						
12 in/45 1893	44.3	44.1/45	750	196	BM 13	2559
12 in/1893–6	40	47.3	750	245		2674
10.8 in/1893–6	40	35.1	562	$187\frac{1}{2}$	BM 13	2674
German guns						
11 in/SKL/40	37	44.6	529	$145\frac{1}{2}$	RPC/100	2690

In the 'charge' column, NC is nitrocellulose: BM is the French nitrocellulose propellant, the numerical suffix indicating a particular strip thickness; 'RPC/100' is the German tubular nitroglycerine/nitrocellulose propellant of 1900.
★It was originally intended to use a 286 lb charge of a larger size MD cordite (55) in this gun, but there were manufacturing difficulties with this size, and the lighter charges of size 45 were used instead.

fewer and longer hoops. This particular gun is noted for the use of a brass cartridge case for the charge. This feature, which was henceforward standard German practice, had not previously been used in a heavy gun mounted afloat, though it was present in the German 9.4 in/57 calibre introduced in the *Kaiser* class. The cartridge case provided the necessary obturation seal for the wedge breech, which in this gun had a flat and not cylindro-prismatic wedge, and the Broadwell ring was abolished.

In British 6 in guns the brass cartridge case was abandoned in favour of the usual silk BL cartridge in the Mark VII gun introduced in the *Formidable*. The other QF features were retained with better breech mechanisms and the Welin breechblock. It should be noted that the German navy retained brass cases for secondary guns.

The 6 in BL Mark VII was 44.9 calibres long and weighed 7.4 tons. With a 100 lb shell and the original charge of 20 lb cordite or 23 lb MD cordite, muzzle velocity was 2536 ft/sec, but with strengthened mountings a charge of

28.6 lb MD was later used, increasing the muzzle velocity to 2762 ft/sec.

The principal gun for defence against torpedo craft, with the exception of the French and German navies, was the 3 in gun usually known as 12 or 14 pounder, which replaced the earlier 6 and 3 pounder guns, though the latter was often retained for use in such positions as the tops or on the turret crowns. A new higher velocity Vickers 3 pounder was in fact introduced in the *King Edward* class. The 3 in was first mounted as an anti-torpedo gun in *Renown* in the British Navy. In Germany 3.4 in guns were retained in all ships, while in France the 3 pounder was still used.

The gun mounted in British ships, from the *Renown* to the first five *King Edwards*, usually known as the 12 pounder/12 cwt, was an all-steel 40 calibre gun firing a $12\frac{1}{2}$ lb shell at 2260 ft/sec. An Armstrong coned breechblock was used, with the thread interruptions each divided into three staggered sections. This gun was also used by Japan and Italy.

There were two types of 8 mm (3.46 in) QF in German battleships. The *Kaiser* and *Wittelsbach* classes had the 27 calibre gun, introduced in the previous period, while the *Braunschweig* and *Deutschland* classes had a 32 calibre gun. This had a flat, instead of cylindro-prismatic, wedge breech. Both guns used a 15.4 lb shell and, with smokeless propellant, the muzzle velocities were 2198 and 2526 ft/sec respectively.

All these guns had an adequate muzzle velocity for action at 1000 yards, but the 12 pounder/12 cwt and the earlier German 3.4 in, suffered at longer ranges from the same danger space against low targets; none of them fired a large enough shell for effective results against destroyers. The increase in power of anti-destroyer guns was to be one of the principal features of the next 10–15 years.

Ballistics in the Black Powder Era

A cursory examination of technical factors influencing the design of ordnance and of the emergence of ballistics as an applied science

J F GUILMARTIN JR

The appearance of the first primitive cannon in Europe in about 1300, marked a major watershed in the history of technology. Gunpowder weapons were the first practical machines capable of converting an inanimate source of energy into power on demand, and it was in Europe that their potential would be realised. That source of energy was gunpowder or black powder, a mixture of saltpetre (KNO_3), charcoal and sulphur, finely ground and mixed together. There is reason to believe that the very earliest gunpowder in Europe was mixed to a ratio of 1/1/1 by weight, and there is evidence that gunpowder of these proportions can expel a projectile from a gun.[1] Practitioners quickly realised that the optimum proportion of saltpetre was much higher, however, and at an early stage recipes began to approximate a ratio of 75/15/10, generally accepted as the modern proportions (Collado, 1592: 24–25. 77–9; H.M. War Office, 1905: 1; U.S. Army, 1964: 15; Harris, Lannon *et al.* 1978: 354, 365).

The issues are far more complex than these introductory comments imply, but this ratio approximates the optimum and provides a convenient datum for general discussion. When ignited, these ingredients produce considerable amounts of gas and heat, through the release of the oxygen resident in saltpetre and its combination with the carbon in charcoal. The substance was first discovered in China, and there is convincing evidence that the Chinese were using it in cannon by the first half of the twelfth century (Gwei Djen, Needham and Chi-Hsing, 1988: 594, 602). However, Chinese gunpowder technology stagnated. By the end of the fourteenth century

Europe was on a par with China in this respect and by the mid-fifteenth century had a clear lead which it would not relinquish. The reasons for Chinese failure to develop more fully the potential of their discovery, an intriguing and important question, is not at issue here. Nor is the means by which gunpowder migrated from China to Europe, as it almost certainly did.[2]

Our concern is with black powder itself, with the technological implications of its development and its exploitation — an area where better understanding can usefully inform research into the history of ordnance as a whole. Scientific analysis of the performance of guns, propellants and projectiles can tell us a great deal about their technical development and operational use, but the issues are not as simple as they were once assumed to be. Experimental ballistics, born as an applied science in the last half of the eighteenth century, had turned away from black powder by the end of the nineteenth century. Black powder's behaviour was clearly understood at that point, and for the relatively undemanding applications, where it could not be replaced by the new and more powerful nitrocellulose-based propellants, principally as booster charges in very large artillery shells, empirical understanding was sufficient.[3] Consequently, experimentation into the behaviour of black powder effectively ceased. At the same time, the array of measuring and analytical tools available to the experimental ballistician was expanding dramatically. In the mid-19th century experimental ballistics was transformed by a host of developments including: an electrical means of measuring short intervals of time with unprecedented precision; the advent of aerodynamics as an applied science and wind-tunnel spark photography. The development of the piezo-electric pressure gauge shortly after World War I, permitting continuous recordings of pressure as a function of time at a single point within a gun, completed the break with the past (U.S. Army, 1964: 67). At the same time, cast steel barrels, hydro-pneumatic recoil mechanisms, reliable fuses and high explosive fillers for shells, optical ranging and sighting devices, and dramatic improvements in manufacturing tolerances combined to change the face of artillery.

In external ballistics — the study of the movement of projectiles in flight — and in terminal ballistics — the study of the interaction between projectile and target — the accumulation of knowledge proceeded in a continuous fashion. However in internal ballistics — the study of the behaviour of charge and projectile within the gun — there was a sharp break with the past. The ballistic behaviour of black powder was sufficiently different from that of nitrocellulose-based, smokeless propellants, that experimental ballisticians had to start afresh. While the developments in instrumentation and artillery technology, alluded to above, did not the cause the break they amplified its sharpness. The answers that the pioneer ballisticians of the black powder era had struggled to obtain were no longer germane to the concerns of the day and their methods had been bypassed by advancing technology. Consequently, their achievements and data have been largely forgotten.

Perhaps overwhelmed by the massive scientific and technical knowledge base underpinning contemporary ballistics, historians have tended to approach black-powder ordnance with the implicit assumption that black and nitrocellulose powders behave similarly, except for the greater power, the lack of smoke and the clean burning properties of the newer propellants. This tendency was no doubt reinforced by a general lack of interest in black

powder among technical experts, who all too often overlooked it as archaic and undeserving of serious attention. The belief that black and smokeless propellants behave similarly has distorted our understanding of the development of ordnance and tactics, notably in fostering the mistaken notion that muzzle velocities of black-powder artillery could be increased more or less indefinitely by increasing barrel lengths. The notion, and its corollaries in external and terminal ballistics, that long range accuracy was commonly attainable with smooth-bore black-powder cannon and that long-range fire was tactically effective, have left their most prominent mark in the debate over the importance of the length of cannon barrels in the defeat of the Spanish Armada.[4]

In fact, black powder and nitrocellulose-based propellants behave very differently. It follows that the design of black-powder ordnance cannot be properly evaluated by applying criteria derived from the study of nitrocellulose-based propellants. Moreover, this article is intended as a first step towards laying down criteria for the analysis of black-powder ordnance on its own terms. In one sense this is a matter of rediscovery, for those criteria exist, or rather existed, having been developed over the centuries at considerable cost. Until the mid-18th century, the process was almost entirely empirical and our analysis must be largely speculative. But with the practical application of science to ballistics (Benjamin Robins' application of Newtonian mechanics to the problem is generally regarded as the watershed), the advance of ballistic knowledge was increasingly recorded in formal fashion, albeit often in technical and scientific terms unfamiliar to the modern reader. The birth of ballistics as an applied science coincided with the end of the black-powder era and much good data was recorded, particularly towards the end of the nineteenth century. While I have made no attempt at an international survey of the technical literature of black-powder ballistics, it is clear that an explosion of practically-applicable ballistic knowledge occurred in North America and Great Britain from the mid-19th century, and that American and British theory and practice corresponded with standards elsewhere. The efforts of Thomas Jefferson Rodman in the United States and of Andrew Noble and Frederick Able in Great Britain, are generally regarded as seminal and I have depended heavily on their work.

I will begin by summarizing the salient differences in behaviour between black and nitrocellulose-based propellants, and briefly outline the practical impact of those differences on ordnance design. I will then assess the nature of the technical challenge that black powder posed for the medieval smith, addressing in the process the ultimate capabilities of black powder as a propellant. From that point I will concentrate on internal ballistics, addressing in parallel the ballistic characteristics of black powder and the emergence of ballistics as an applied science.

My focus will be on artillery, particularly heavy ordnance. I have adopted this viewpoint for three reasons. First, the internal ballistics of small arms form a discrete and important subject, which merits attention in its own right. Second, the design and manufacture of heavy ordnance formed the cutting edge of technological advance in weapons' design from the beginning. Indeed, if the relevant criteria are maximum temperatures, pressures, velocities, and rates of power production, one can reasonably argue that heavy ordnance marked the leading edge of technological advance as a whole, until the advent of jet engines and rockets capable of lifting payloads into space. Third, since the design of heavy ordnance was both technologically demanding and important to the state, particular effort was devoted to amassing relevant data. This was particularly true towards the end of the black-powder era, as the link between science and technology became explicit, and much of that data survives.

The essential differences between black and smokeless propellants can be quickly summed up. The propellants that replaced black powder are solid, barely stable chemical compounds, whose large constituent molecules contain oxygen, hydrogen, carbon and nitrogen. The large molecules break apart when ignited and their constituent atoms re-combine into many smaller gas molecules, producing large quantities of gas and heat (U.S. Army, 1964: 9). The salient ballistic characteristic of these propellants is the positive relationship between burning rate, pressure and temperature; the higher the pressure and temperature within the gun, the higher the rate at which propellant gasses are evolved. Black powder, by contrast, is a simple mixture rather than a chemical compound and produces propellant gasses by means of a complex series of chemical and thermodynamic reactions, which remain poorly understood.

Black powder has three salient ballistic characteristics: first, it decomposes by means of three mutually dependent, chemical reactions, one being exothermal, the other two endothermal. These reactions proceed in parallel, cancelling one another out thermodynamically and rendering the rate of propellant gas evolution practically independent of changes in pressure and temperature. These reactions are approximations of a more complex chemical and thermodynamic reality although that reality does not appear to depend on the precise ratio of ingredients or the physical characteristics of the powder. Seemingly large variations in the percentage of potassium nitrate, (saltpetre or KNO_3) have a surprisingly modest effect on performance, and the sulphur can be dispensed with altogether, at least for powder used in large ordnance (Noble, 1907: 6). In short, the net decomposition reaction has surprisingly robust mechanisms to compensate for changes in composition. This evidence is supported by the fact that powders which produce more heat produce less gas and vice versa (Noble, 1907: 6).

Second, the ballistic performance of black powder is critically affected by its physical form. The earliest gunpowder, called serpentine, was compounded dry of ingredients, ground separately to a fine powder and mixed together. Later, the ground ingredients were mixed with water and then dried. Saltpetre being highly hygroscopic bound the components together permanently within the fibrous structure of the charcoal. The powder then formed grains which segregated according to size. The result, called corned powder, was reputed to be more powerful than serpentine. This argument might be debated on theoretical grounds, as indeed it was, but the point is academic. Corned powder was clearly more uniform and predictable in performance and by the mid-16th century was considered dangerous when used in the older, wrought iron pieces. This was almost surely the case, though we can only speculate as to why.

Later, gunpowder was tumbled in drums to wear the rough edges off the grains and to impart a glaze, making the powder more stable and the grains more uniform. Apparently glazing retards the burning rate, though this was probably not appreciated until the nineteenth century. Eventually small amounts of graphite were added during tumbling to enhance both the glaze and the retarding effect (H.M. War Office, 1895: 45). From the mid-nineteenth century, large-grained artillery powder was formed in hydraulic presses which controlled precisely both the shape and the density of the grain.

The reasons for these important changes in powder manufacture are simple; all changes affected the critical determinants of black powder's ballistic performance: its grain size, shape and density. Once a low threshold pressure of about 25–30 atmospheres (370–440 psi) is attained, the burning rate within a grain of black powder is about 0.20 fps (0.06 m/sec), while the rate of propagation from grain to grain is some 30 fps (9 m/sec), over two orders of magnitude higher (Wright, 1985: 2, 6; Blackwood and Bowden, 1952: 304;

Harris, Lannon, et al. 1978: 369). Since burning occurs at the surface of the grain and the burning rate is constant, the rate of gas evolution is a function of surface area. Since surface area varies inversely with grain size, charges of smaller-grained powder evolve their gasses more rapidly. With spherical grains, the net surface area, and hence the rate of gas evolution, diminishes geometrically once the charge is fully ignited.

In the final years of the black powder era, powder for heavy ordnance was made in the shape of large spheroids, large rounded-off cubes and, finally, large, perforated cylinders and hexagons. Logic suggests that more densely compacted powders burn more slowly although, to my knowledge this has not been tested, and late nineteenth century ballisticians were of this opinion (H.M. War Office, 1905: 1). British prismatic powder, pressed into flat-sided hexagons 1 inch thick and 1.4 inches across from face to face, perforated with seven 0.2 inch holes (Noble, 1871: 6) was probably the most sophisticated of these. The surface area of a burning hexagonal plate diminishes less rapidly than that of a sphere, and burning within the holes increased surface area as the charge burned. This geometry probably promoted so-called progressive burning through a net increase in surface area as the grain burned (this is a common characteristic of modern smokeless artillery propellants) and, in combination with the large grain size, was unquestionably effective in slowing the initial rate of gas evolution and sustaining a higher average pressure within the bore thereafter.

The extent to which these generalisations about grain size and geometry apply to serpentine powder is unclear. Was a charge of serpentine in effect a single grain of infinite size, an infinite number of grains of zero size, something between these two extremes, or something else altogether? The answer is that we simply do not know, but — as is usual with black powder — the more complex possibilities seem the more likely.

Third, about only 43% of the decomposition products of black powder by weight are gaseous, the other 57% being solid (H.M. War Office, 1905: 29). By contrast virtually all the decomposition products of modern gun propellants are gaseous, small amounts of water vapor being the main exception. The high molecular weights of the decomposition products of black powder, combined with the constant burning rate, places an absolute upper limit of about 2,500 fps on velocities attainable from guns. Further, the molecular mass of the decomposition products combined with the fixed rate of gas production places an absolute limit on the distance over which a black powder charge can continue to impart acceleration to a projectile. The optimum length of bore for heavy ordnance has traditionally been expressed in calibres, or multiples of the bore diameter, and many writers, this author among them, have accepted the traditional wisdom that barrel length, in calibres, was a primary determinant of ballistic performance irrespective of bore diameter. This view, however, embodies inconsistencies. By the 16th century, if not before, empirical wisdom indicated that optimum barrel lengths in calibres for small arms, were far greater than for heavy ordnance. Moreover, there seems to have been an implicit awareness, evident by the late 17th century and clearly apparent by the mid-18th, that optimum barrel lengths in calibres for heavy artillery were shorter than for lighter pieces.[6] Thermodynamic considerations come into play, for bores of greater diameter conserve heat more efficiently, a point offered by Rodman as a possible explanation of the greater pressure developed in larger guns by equal weights of powder behind equal weights of shot.[7] Since larger bores waste less heat, they presumably develop the same velocities over a shorter distance, all else being equal. However, this line of argument, borrowed from calculations of exhaust gas velocity and propulsive efficiency in rocketry, suggests that the limiting factor is not the relative length of the bore, but the molecular weight of the decomposition products.[8] The critical parameter, in short, is not the relative length of the bore, but the distance which the evolved decomposition products can travel from their point of origin whilst still exerting a meaningful increment of propulsive force.

The value of that distance is probably beyond theoretical calculation, though British experiments at the very end of the black powder era give some idea of its ultimate limits. Tests in a 6 in rifled naval gun using a 50 lb charge of brown prismatic (a very large-grained, slow burning powder of the type mentioned above), showed continued acceleration up to 47 ft or 100 calibres, but 90% of the velocity was attained within 21.6 ft or 50 calibres. Further, since the charge was very large and was ignited from the rear, it no doubt moved forward as it burned, thus extending the effective barrel length. In the same test series a 23 lb charge of RLG_2, a relatively fine-grained, fast-burning cannon powder, imparted 80% of the final velocity to the shot in 10 ft, and 90% in 16.5 ft, or 40 calibres (Noble, 1900: 6, Fig. II). This is probably close to the limit, and the maximum effective barrel lengths of ordinary service cannon firing charges of relatively fine, spherically grained powder were much shorter. American experiments with a twelve pound smoothbore gun in the 1850s showed that the maximum velocity with a normal service powder charge was obtained with a barrel of 25 calibres, or 25 times the bore diameter, about 9 feet 8 inches (2.9 m), including the portion of the bore occupied by the powder charge. However, the same tests showed that extending the bore beyond 16 calibres, or some 6 ft 3 in (1.9 m), resulted in an increase in velocity of only 5.5%. Taking into account the space taken up by charge and wad (about a calibre and a half based on contemporary drawings), this suggests that a charge of relatively fine, spherically grained black powder from a fourth to a third of the weight of the projectile, fired behind a cast iron ball in a smoothbore cannon, imparted some 95% of the attainable velocity to the ball by the time it had travelled eight to ten feet from the face of the charge. A 19th century British experiment with a smoothbore gun 20 calibres in length, of unspecified size (though a 24 or 36 pounder seems most likely from internal evidence), was fired to determine the effect on velocity of progressively increasing the powder charge. Tests showed that the velocity increased until the charge occupied $\frac{1}{4}$ of the bore and equalled $\frac{7}{8}$ of the weight of the shot. Calculations show that if the gun in question were a 24 pounder, the maximum velocity was imparted to the shot at a distance of 7 ft 8 in (2.3 m) from the face of the charge, and that if the gun were a 36 pounder the distance was about 8 ft 6 in (2.6 m).

The coincidence in these values is suggestive and by the end of the black powder era, naval ordnance, at least, was cast in approximation of them: the main U.S., British and French naval guns in service in 1856, ranging from long 32 pounders to 10 in pivot guns, varied in absolute length of bore from 8 ft 4$\frac{1}{3}$ in (2.24 m; the U.S. 8 in shell gun) to 9 ft 4$\frac{1}{4}$ in (2.85 m; the U.S. 64 pounder). Subtracting 1$\frac{1}{2}$ calibres for charge and wad, the bores range from 7 ft 4$\frac{1}{8}$ in (2.24 m; the U.S. 8 in shell gun, with the U.S. 10 in gun close at 7 ft 7 in, 2.31 m) to 9 ft 4$\frac{1}{4}$ in (2.85 m; the U.S. 64 pounder) (Tucker, 1989: 197; bore length is not given for the British 68 pounder).

On balance, then, a figure of eight to ten ft from the face of the charge to the muzzle seems a reasonable rule of thumb for the maximum effective barrel length of smooth bore black powder ordnance, certainly for 12 pounders and above.

This hypothesis, however, is based on 19th century data and we are also concerned with earlier periods. Early powders were less dense and charges larger, amounting in the 16th century to the weight of the ball for culverins. Such a charge might have occupied as much as four calibres of the bore and it is reasonable to speculate that such charges might have increased optimum barrel lengths.[9] The evidence suggests that if they did so, it was not by much. A 19th century French test of a 36 pound cannon, 16 calibres long with a powder charge equal to the weight of the ball, produced a muzzle velocity of 1,320 fps, while a 42 pound charge produced only 1,170 fps and additional increments of powder reduced the velocity (Benton, 1862: 130). Thus, while we cannot be certain, particularly where serpentine powder is concerned, it seems unlikely that the very large powder charges characteristic of the 16th century and earlier extended optimum barrel lengths significantly in either relative or absolute terms.

Having roughly delineated the limits of black powder as a source of power, we must touch on the effect of those limits on the process of engineering and design. Until the advent of gunpowder weapons, the cutting edge of power production lay in developing and harnessing animate energy — particularly horse power — and in capturing the energy resident in natural forces, notably wind and falling water. Gunpowder represented a sharp break with the past, for it operated in ways which the human senses could not perceive. Horses, wind and water were all capable of producing forces which no human could resist, but their action could be observed directly and studied in a straightforward manner. Gunpowder, by contrast, worked with a swiftness, heat and violence which made direct observation impossible, a problem compounded by the consideration that it behaved diferently, or seemingly so, in an enclosed space. The smiths who harnessed gunpowder thus proceeded very differently from those who exploited the energy resident in muscles, wind and water. The closest analogs to their problems were those confronted by metal-workers, particularly those who smelted metal from ores in enclosed, forced-draught furnaces. Indeed, the coincidence in time between the appearance in Europe of cannon and the Catalán forge, which marked a revolution in iron founding, is suggestive, hinting at common underlying economic, social and political driving forces.[10] But those who smelted ores and cast metals were concerned with the transformation of materials over measurable intervals of time, not with the containment of forces generated seemingly instantaneously.

The difference in the magnitude of the problems with which the two had to deal, is thrown into stark relief by comparing extreme physical parameters: Bronze melts at around 1066°C (1950°F), molten iron emerges from a simple forced-drought furnace at temperatures between 1260° and 1540°C (2300°–2800°F) (Kent, 1938: 20–11; 4–26)[11] and metals were smelted, cast and forged at ambient atmospheric pressures.[11] By contrast, the temperature of explosion of black powder is about 2200°C (4000°F) (Nobel and Able, 1874: 12; Noble, 1909: 12) and a quantity of black powder exploded within its original volume produces pressures in the order of 6,500 atmospheres (94,100 psi; 6.6×10^7 kg/m^2) (Noble, 197: 3).

In fact, even the most efficient gun can capture only a faction of the energy suggested by these values. British experiments in the 1870s recorded steady state chamber pressures in cannon as high as 4,270 atmospheres (62,700 psi: 4.5×10^7 kg/m^2); this figure, just over two thirds the theoretical maximum, probably represents the practical limit of black powder's propulsive capacity.[12] Such theoretical inefficiency, however, had little practical significance, for the discovery of black powder pushed propellant technology well beyond the capabilities of contemporary structures and materials. The problem was not how to develop the energy resident in black powder, but how to contain it, a problem exacerbated by the fiendish unpredictability of the stuff. While net pressures produced in guns were well below the theoretical maximum, transient 'spikes' routinely exceeded it and presures as high as 9,660 atmospheres (142,000 psi; 9.96×10^7 kg/m^2) were reliably recorded (Noble 1907: 4). Due to scale effect, the relatively greater strength of smaller structures as

an inverse function of size, and their thermodynamic inefficiency, small arms were relatively safe from all but the most blatant abuse such as firing with a clogged muzzle. But cannon, particularly large cannon, were a different matter and it is probably safe to assert that there was always a powder charge on hand capable of destroying the most strongly constructed of black powder cannon. In short, to harness even a fraction of the energy resident in black powder, medieval smiths — engineers in reality, if not by formal certification — had to deal with unprecedented pressures and temperatures, developed over unprecedentedly short periods of time.

The record of how they did so remains only in a meagre handful of illustrations, archaeological artifacts and archival notations. We can only guess at how they turned the trick, but turn it they did, and in so doing changed the smith to an engineer. The anonymous smiths who forged the guns of the *Milamete* manuscript were followed by others who doggedly applied their hard-earnt knowledge to the same problem, how to convert the energy resident in saltpetre, charcoal and sulphur into destructive power. The process of design evolution was almost purely empirical, using destructive testing — intentional or otherwise — to assess incremental changes in materials and form. This methodology hardly fits modern notions of scientific design, but crude as it was it was remarkably sophisticated. We cannot yet evaluate the design of medieval and early modern cannon except in terms of operational success or failure, for we know little about the materials used, and the required stress calculations are extremely complex. We do know that at least the better founders, and later foundries, followed precedent with precision and care, systematically advancing the art by incremental changes to accepted models (Guilmartin, 1974: 172–74, 284–91; Guilmartin, 1983: 579–80). We also know that over time they produced lighter, safer cannon, though the rate of change was for many centuries very slow.

Nor were the consequences of their efforts negligible. Cannon were the first practical internal combustion engines, albeit with throw-away pistons and a destructive purpose, and were to remain the only ones until the advent of the gasoline engine. Of more consequence, European success in harnessing the next major inanimate source of energy, steam, owed much to gunpowder technology. Practical steam engines could not be built until cylinders and pistons could be machined to precise dimensions with replicable tolerances. When James Watt arrived at this conclusion, the requisite technology was already on hand, developed for boring cannon (Trebilcock, 1969: 477, cited in McNeill, 1982: 212). Nor is this link between the exacting technical demands of heavy ordnance manufacture and technological advance as a whole an isolated example: one author has credited the pioneering British artillery designer and manufacturer Joseph Whitworth with 'rendering engineering a precise science' (Trebilcock, 1969: 479). It is worth noting that Whitworth became professionally associated with Andrew Noble, the source of much of our data in 1863, through the amalgamation of his ordnance concern with that of William Armstrong, in which Noble was a partner.

The struggle to master gunpowder thus has an interest and significance well beyond the narrow concerns of military or naval history. This is particularly true in tracing the development of the connections between science and technology, for it may reasonably be argued that it was through internal ballistics that applied science in the modern sense first began to affect the design process in a meaningful way. Our concern is with the properties and characteristics of black powder rather than with the design of guns, but the two cannot be cleanly separated and if we are to assign priority to the one I would argue for gunpowder, for if we are to understand the history of ordnance design, we must understand the nature of the forces the designer sought to control and how knowledge of them evolved with time.

Just as internal ballistics cannot be separated from the design of ordnance, it cannot be meaningfully addressed without considering the contribution of external and terminal ballistics to the operational and intellectual context within which it developed. Guns were made to hit targets, and the fall of the projectile had to be predictable if the effort expended in expelling it were not to be wasted. Much verbiage was expended on trajectories by Tartagila and other early writers on ballistics, but their speculations lacked experimental foundation and the range tables which accompanied many of their works were fanciful in whole or in part (Hall, 1983: esp. 111–17; Tartaglia, 1588: 12–24). Galileo demonstrated that the ballistic trajectory of a projectile in a vacuum was a parabola, and from the mid-18th century, if not before, it was understood that aerodynamic drag represented the difference between the theoretical and actuality. The first effective application of science to the problem in the modern, experimental sense, however, came only after 1740 with Benjamin Robins' development of the ballistic pendulum ('Chronograph', 1910: 84; Muller, 1973; 212, Fig. 10). This permitted measurement of the velocity of a bullet through the straightforward application of Newtonian mechanics. A projectile was fired against a massive metal pendulum, transferring the projectile's linear momentum to the pendulum as angular momentum; since the mass of the pendulum was known and its swing could be measured directly, calculating the impact velocity of the projectile was a straightforward matter ('Chronograph', 1910: 84). The results showed that theorists had badly underestimated the effects of aerodynamic drag, but to little practical effect. Aerodynamic drag proved intractable to precise mathematical description and the behaviour of ballistic projectiles in flight could not be described with militarily useful precision until relatively recently.[13] By the end of the black powder era, theory sufficed to provide first order approximations of the behaviour of projectiles in flight, but theory and mathematics were used in constructing gunners' range tables only for interpolation between known, experimentally-determined values. In the words of an eminent ballistician writing in the 1911 *Encyclopaedia Britannica*:

> *The theoretical assumptions of Newton and Euler of a resistance varying as some simple power of the velocity, for instance as the square or cube of the velocity (the quadratic or cubic law), lead to results of great analytical complexity, and are useful only for provisional extrapolation at high or low velocity pending further experiment (Greenhill, 1911: 271).*

In so far as artillery was concerned, practical limitations on accuracy rendered these aerodynamic and mathematical difficulties largely irrelevant, even following the widespread adoption of rifled artillery and effective explosive shells in the second half of the 19th century.

European gun founders abandoned the arrow-like projectiles of the *Milamete* manuscript at an early date and adopted spherical projectiles fired from tubular barrels, a combination which, though efficient in converting gunpowder's energy to destructive power, was inherently inaccurate. To prevent the projectile from sticking in the barrel, clearance, or windage, between ball and bore was necessary. This was required to allow for irregularities and inaccuracies in ball and bore and because firing built up layers of fouling which could only be removed by laborious cleaning; in addition, cast iron shot expand as they rust, a particular problem when operating near salt water. The first of these problems could be reduced by improved manufacturing techniques, but the second two were inherent to black powder smooth bore ordnance. Collado, writing in the 1570s, said that the ball should be 10% smaller than the bore, that is 10%

lighter than a ball equal in diameter to the bore (Collado, 1592: 43–4, 48). This works out to about $3\frac{1}{2}\%$ of the bore diameter. The general rule for British naval ordnance was a twentieth of the bore by the mid-18th century and remained so through the early 19th, though carronades had less windage; this seems to have been considered relatively generous and contemporary American naval practice was one twenty fifth of the bore. There was a general reduction in windage with the adoption of improved casting and machining techniques and in 1840 the US Navy set windage at between 0.10 and 0.20 in for all bore diameters (Tucker, 1989: 147). The contemporary French standard was 0.133 in and at leastt one British authority considered that anything in excess of 0.15 in (the standard for carronades was 0.14 in) was undesirable for 12 pounders and above (Douglas, 1855: 99).

Windage permitted the ball to rebound from side to side, or ballot, as it travelled down the bore, departing at an unpredictable angle. Whatever spin the ball acquired in the process was at an angle to the long axis of bore, causing the ball to deviate from the trajectory dictated by velocity, mass and aerodynamic drag along the axis of flight (Douglas, 1855; 86–7). The amount of spin imparted to the ball seems to have been relatively minor, and was eliminated in the 19th century by the adoption of sabots, cylindrical wooden blocks strapped to the rear of the shot. But the root of the problem was more basic; non-spinning or slowly spinning spherical projectiles are inherently inaccurate aerodynamically, having a tendency to 'float' or wobble, deviating unpredictably from the line of flight.

The problem of inherent inaccuracy was overcome in principle with the adoption of rifled artillery in the last half of the 19th century, but rifled ordnance did little to improve accuracy practically attainable in the field or at sea until the appearance of hydro-pneumatic recoil mechanisms at the end of the 19th century. Until that point, most artillery carriages consisted of rigid beds of timber with iron fittings which differed little in their essential characteristics from those of the late 15th century. These recoiled in erratic fashion, throwing the point of aim off before the projectile left the muzzle, frustrating attempts to take advantage of rifled artillery's inherent accuracy. Cast iron carriages offered some advantages, but were deemed too heavy for field and naval use. In the 1880s and 90s, long range accuracy was obtained with rifled black powder fortress and naval ordnance which approached current standards, at least under ideal conditions (Noble, 1890: 11). This remarkable achievement, and the technology which made it possible, had important implications, paving the way for the developments which revolutionized gunnery and ordnance design and manufacture in the 20th century and in many ways anticipating them. The ordnance in question, however, was highly specialized and in the event had little operational impact. While an important part of the history of ordnance and deserving of scholarly attention in its own right it is beyond our concerns here.

Practical gunners understood the issues involved in striving for accuracy long before the advent of rifled artillery. In listing the causes of inaccuracy, the 16th century Spanish gunner Luis Collado emphasised those which took effect during recoil, such as rocks behind wheels, wheels loose on their axles and imperfections in the carriage (Collado, 1592: 38). But given the available structural materials for carriages, there was little that could be done to enhance accuracy and little benefit in so doing given the limited terminal effect of an inert iron ball at long range. Siege and fortress guns, which could be carefully laid for fire at known ranges, were a special case and here 19th century advances in external and terminal ballistics had some effect, for siege and garrison carriages did not have to be kept light for service in the field and could be fitted with sliding mounts which controlled recoil far better

than field carriages. This was also true of naval ordnance and the sliding mounts of late 19th century shipboard pivot guns, with their friction recoil mechanisms marked an apex of sophistication in field and naval carriage design prior to the advent of hydro-pneumatic recoil, though wave movement largely cancelled out the advantage in most situations. In the terms of our overview, the effect of advances in external ballistics on accuracy practically attainable in the field or at sea, expressed most cogently in maximum effective ranges, were negligible during the black powder era. By the time reliable time-fuzed explosive shells, 'soft' recoil and rapid pointing mechanisms, training and aiming devices were developed, the age of smokeless propellants was at hand.

This is not to say that black powder gunners could not hit anything, that long range accuracy was unattainable and tactically irrelevant or that standards of accuracy did not improve with time. By the 16th century, exceptional gunners, given a good, well-mounted piece and the opportunity to carefully lay it, were capable of effective counter-battery fire at ranges approaching 1,000 yards (925 m) where they could see their target. The point is made by a particularly well-documented incident at the siege of Siena in early 1555 in which a Sienese demi-cannon fired by a skilled gunner did such damage to the main imperial battery that the imperialists were forced to shift their efforts to another sector (Pepper and Adams, 1986: 135–37). The same example also effectively illustrates the practical limits of accuracy and terminal effect: the imperialist battery was sited about 225 yards (200 m) from the section of wall it was emplaced to batter, probably about the maximum range at which battery could be expected to have much effect. Long range artillery fire was even less common to field engagements than to sieges, for it was expensive and difficult to carry munitions into the field and firing cast iron shot against bodies of troops at extreme range was generally a waste of effort.

The problem of accuracy in naval gunnery was more complex, particularly where seagoing ships with broadside ordnance are concerned. The roll of the ship combined with inherent limitations in accuracy to restrict most naval gunfire to relatively short ranges of 300–500 yards (275–450 m) or less. The potential benefits of a long shot which disabled a hostile ship ensured that long range shooting was undertaken on occasion, but, as with long range gunnery on land, shooting at extreme ranges was the exception rather than the rule.

Still, by the mid-16th century it was apparent to at least some European naval commanders and gunners that the ability to drive heavy shot repeatedly against an enemy hull represented an enormous tactical advantage. Many practical difficulties stood in the way of realising this ideal, carriage design among them, and I agree with Colin Martin and Geoffrey Parker that the development of the four-wheeled naval truck carriage before the middle of the 16th century was an epochal development (Parker, 1988: 95–6; Martin and Parker, 1988: 50–51). A truck carriage on the deck of a ship was a far better recoil absorption mechanism than contemporary field carriages, and was hence inherently more accurate. In its developed form, the truck carriage was also easier to handle and position accurately. The roll of the ship remained a problem, for only the smallest swivel guns could be trained quickly enough to negate it, but roll could be compensated for at moderate ranges by careful timing supported by quick and reliable ignition. The problem of quick and reliable ignition was peculiarly important to naval gunnery, and it is worth noting that the application of linstock to touchhole was deemed to have important advantages over flintlock and percussion ignition mechanisms to the very end of the smoothbore black powder era. Higher muzzle velocities would also have helped by reducing the time of flight of the shot, by reducing the magnitude of elevation adjustments needed to compensate for changes in range, and by reducing the adverse effect of

range estimation errors by flattening the trajectory. These factors were understood by the final decades of black powder, smoothbore naval ordnance and, indeed, hotly debated.[14] It should be noted, however, that the debate hinged on differences of a very few feet and a very few seconds and took place within the context of a knowledge and technology base that was the product of centuries of incremental development. By the nineteenth century naval ordnance and artillery practice had reached a sufficiently high level of standardization and sophistication that such small differences could be appreciated and consciously exploited. The extent to which this was the case previously remains an open question, but it seems unlikely to me that such differences in muzzle velocity as existed among naval ordnance had any real effect on practical accuracy before the mid-eighteenth century at the earliest, though one author has argued that they probably did in the 1588 Armada campaign (Padfield, 1988: 88).

The contrast between artillery and small arms is instructive in considering the effect of the general adoption of rifled weaponry during the second half of the nineteenth century. The body of a trained marksman is a highly efficient recoil control mechanism and the accuracy of rifled black powder infantry weapons was excellent, extending out to 300–500 yards (275–450 m) or more for trained riflemen engaging formed bodies of troops. The development of the expanding-base conical bullet by the French captain Minié made the rifled musket a practical weapon for general issue, extending the effective range of infantry fire beyond that of field artillery for a time. This accuracy, however, was achieved by empirical means. In short, it is safe to say that while the study of external ballistics, the behaviour of projectiles in flight, stimulated much scientific thought during the black powder era it had little practical impact on the design and construction of ordnance until almost its end.

The study of terminal ballistics, too, was undertaken in an almost entirely empirical fashion until the very close of the black powder era when the appearance of armored warships forced a burst of experimentation. That is not to say that medieval and early modern gunners had no understanding of the basic parameters involved. Indeed, in a rough sense they may be said to have expanded their knowledge by scientific means, regarding each naval engagement or battery of a fortress as a more or less replicable experiment. By the 16th century practical soldiers and sailors understood the differing effects of projectiles of stone, cast iron and lead on various targets and employed appropriate kinds of ordnance accordingly. Sixteenth century gunners consciously used the greater penetration of iron projectiles in conjunction with the greater shock effect of stone projectiles on masonry to bring down sections of fortress wall (Pepper and Adams, 1986: 135–6; Collado, 1592: 13). Similarly, gunners understood the respective advantages of solid shot, case and cannister and, when it appeared in practical form, shell. By the mid-1800s, naval gunners had accumulated a substantial body of knowledge concerning the destructive effects of various types of shot and shell against the wooden hulls of warships at various ranges (Douglas, 1855: 122–48).

The history of the carronade, a gun designed to fire the largest possible shot in terms of its own weight at low velocities, provides evidence of the way in which this knowledge was applied. Introduced by the Carron Company in 1778, carronades were extremely short, with bores of 7 to 8 calibres; even the 68 pdr was only 5 ft 2 in (1.6 m) long, of which a bare four feet was barrel, and 18 pdr carronades were only some 4 ft 6 in (1.4 m) long overall (Lavery, 1987: 104–9). Carronades used a powder charge a third to a fourth that of a normal cannon firing the same projectile and took advantage of the fact that the smashing effect of a cannonball striking a wooden hull was enhanced by reducing the impact velocity below that produced by a normal naval gun at short range (Douglas, 1855: 116). Carronades weighed about a fourth as much as normal naval guns firing the same ball, giving a much heavier broadside for the same weight of ordnance. They were, however, explicitly short range guns and it is abundantly clear from the operational record that a carronade-armed ship could be shot to pieces from a distance by a ship armed with long ordnance without being able to reply. Carronades were unusual in that their maximum range and effective range were essentially the same. They were efficient in terms of destructive power per weight of gun because they were designed to have precisely that characteristic, though to phrase it so is to apply inappropriate 20th century terminology to an 18th century problem. The development of the carronade represents applied 18th century ballistics at its best, highly effective for its intended purpose but representing art more than science. Suffice it to say that the mathematics of terminal ballistics are even more forbidding than those of external ballistics and empiricism firmly predominated until the day of black powder was almost over.

For almost the entire black powder era, gunner and cannon founder were concerned first and foremost with what went on inside the gun between charge ignition and exit of the projectile from the muzzle, and it is to this problem that we will devote the bulk of our attention. The earliest stirrings of internal ballistic thought, of which written evidence remains, involve concern over the composition and manufacture of gunpowder. It was understood early on that something bound up in saltpetre — Collado's term was 'occult properties' (Collado, 1592: 77) — was the source of gunpowder's propellant energy. At the same time it was apparent that gunpowder was more difficult to control in large ordnance than in small and the idea soon surfaced that a smaller proportion of saltpetre was preferable in gunpowder for large ordnance. Biringuccio, writing around 1540, gave a ratio of 3/2/1 for heavy guns, $5/1\frac{1}{2}/1$ for medium guns and alternative recipes of 10/1/1 and $13\frac{1}{2}/2/1\frac{1}{2}$ for small arms (Biringuccio, 1942: 413). The high proportion of saltpetre in the latter two recipes may be attributable in part to impurity, for techniques for producing and distilling the stuff were sophisticated and seem to have spread slowly (Williams, 1975: 127). But knowledge of the process seems to have spread during the 16th century and Collado, writing almost a half century later, specified ratios of 5/1/1 or 4/1/1 for artillery and 5/1/1 or 6/1/1 for small arms, the latter recipe approaching the modern optimum (Williams, 1975: 127–30; Collado, 1592: 25, 77–9).

About the utility of the belief that the proportion of saltpetre should be reduced in gunpowder for large ordnance we can only speculate. The last, most sophisticated black powders for heavy guns had more, not less, saltpetre. British powder for large ordnance in the last years of the 19th century had from $77\frac{1}{2}$ to 79% KNO_3 (H.M. War Office, 1905: 1). This increased proportion, however, was based on careful control of the composition of the charcoal (see below) and on experimental findings that very little sulphur was needed in gunpowder for large ordnance (H.M. War Office, 1905: 1; Noble, 1909: 8). It was also undertaken in light of the ability of steel guns to withstand the stresses imposed by more powerful charges, so that we cannot judge medieval powder by modern standards. The variable proportion of sulphur in the early recipes also raises questions. Sulphur reduces the initiation temperature of the decomposition reaction, and this factor may have been critical in the very early stages of gunpowder development. However, this does not account for the enduring popularity of 75/15/10 powder and we are left to speculate. Of gunpowder's constituents, however, it is charcoal which raises the most questions.

Early smiths and gunners believed that the kind of wood from which charcoal was burned significantly affected ballistic performance. Biringuccio and Tartaglia, writing on either side of 1540, expressed a preference for willow charcoal for large ordnance and hazelnut charcoal for small arms (Biringuccio, 1942: 413; Tartaglia, 1588: 72–75). This preference

for willow charcoal for heavy ordnance proved durable[15] and received a measure of experimental verification in the 19th century as did the strong preference which emerged for dogwood charcoal for small arms and sporting powder. British experiments showed that dogwood charcoal burned with saltpetre evolved 24.4% more gas per unit weight than fir, chestnut, hazel or filbert charcoal and that willow charcoal produced 16.6% more (H.M. War Office, 1895: 15). But the quantity of gas evolved per unit weight was only part of the problem: the British War Office publication from which the above data was extracted goes on to say that dogwood charcoal was preferred for:

> . . . all military small-arms powders, as well as for the best sporting gunpowder. It has been found, however, that cannon powders made from dogwood charcoal are, other things being equal, much more violent in action than those manufactured with willow or alder charcoal. Accidents with powder made from dogwood charcoal have usually proved more destructive than those with any other description (H.M. War Office, 1895: 16).

The tone and substance of the document give no reason to doubt the essential accuracy of this statement, but an open question remains. What was it that made gunpowder compounded of dogwood charcoal dangerous in large ordnance? The complex volatile chemicals which typically comprise 12 to 24% of charcoal by weight are the prime suspect. These 'impurities', perhaps acting in combination with the fibrous structure of the charcoal — both surely species-specific — must have played a role in promoting the formation of shock waves within the chamber. These were the cause of the peak pressures noted earlier, and the fact that shock-wave-induced pressure peaks are localized and transient makes them no less destructive.

Engineering theory tells us that strain is proportional to stress, and the maximum

Table 1
The Effect of Overburned Charcoal on Ballistic Performance

	Charcoal	No 1 burnt 7 hours at low heat	No 2 burnt 4 hours at a greater heat	No 3 burnt 3 hours at a very high heat	No 4 burnt $3\frac{1}{2}$ hours at a heat between 2 and 3
Analysis of	C	78.23%	82.23%	87.55	85.57%
charcoal	H	3.67	3.31	2.91	3.02
	O_2 & trace N_2	19.96	13.19	8.29	10.09
	Ash	1.41	1.27	1.25	1.32
Mean muzzle velocity		1417 fps	1399 fps	1353 fps	1403 fps
Pressures	chamber bottom	20.60 tons	15.68 tons	9.62 tons	13.20 tons
	mid-chamber	15.66	12.76	9.66	11.68
	base of shot	14.22	12.18	7.25	10.46

(H.M. War Office, 1895: 16–17)

stress on a gun is a function of the peak pressure to which it is subjected. Charcoal which was burned longer and at higher than normal temperatures, thus reducing the percentage of volatile constituents, was used by preference in the last and most fully-developed black powders for heavy ordnance. Such over-burned charcoal produced less gas per unit volume; for example overburned willow charcoal produced 18% less gas per unit weight than willow charcoal rendered at normal temperatures (H.M. War Office, 1895: 15). Table 1 gives the results of a British War Department test of four lots of cannon powder compounded from charcoal burned to varying degrees, but otherwise identical, gives both a sense of the complexity of the issues involved and a feeling for the state of the art at the time.

Unfortunately, we know neither the size of gun nor the means by which the pressures were determined (though the values are probably peaks rather than averages) and the complex volatile ingredients in the charcoal were reduced to the elemental state in the quantitative analysis, but the outlines of the results are clear. Both muzzle velocities and internal pressures correlate positively with the percentage of volatile ingredients in the charcoal. The relationships, however, are neither simple nor linear. Those responsible

for the test considered charcoal No. 4 '*best adopted for the gunpowder in question, as combining good muzzle velocity with moderate pressures*'; pending further research, we shall have to take them at their word (H.M. War Office, 1895: 16–17).

Recognition of the importance of charcoal's composition to ballistic performance seems to have been accompanied from the beginning by an awareness of the importance of the powder's physical characteristics. The disadvantages of serpentine were numerous and apparent: the ingredients separated out according to density when subjected to vibration, for example during transport, and fast or careless handling raised clouds of noxious and potentially explosive dust. More important, the ballistic performance of serpentine was evidently variable and difficult to control, a point confirmed by modern experimentation (albeit with small charges, Williams, 1974: 117–19; n. 1 above).[16] Wet incorporation solved most of these problems and offered advantages in cost as well since water power could be readily adapted to its manufacture. By late medieval times the ballistic advantages of wet incorporation and the forming of the powder into uniform grains were appreciated, though not understood in a scientific way.

Medieval and early modern gunners had no means of directly investigating what transpired within the gun. However they had some surprisingly accurate ideas about how and why black powder behaved as it did. They recognized that very short barrels were inefficient; in Biringuccio's words . . . *it is obvious that the longer the tube of the gun, the more vigorously and farther the fire sends the ball* (Biringuccio, 1942: 241–2). They also recognised that there was an optimum barrel length, beyond which no added increment of useful velocity could be attained, though they were anything but clear on just what that length was or what determined it (Tartaglia, 1588: 30). More surprisingly, at least the more perspicacious 16th century gunners understood empirically that erratic burning of the charge could produce large and dangerous forces which destroyed accuracy and could lift a large cannon barrel out of its carriage (Callado, 1592: 10). Further, they correctly attributed these forces to improper touchhole placement. To put the problem in modern terms, they empirically understood the problems caused by the formation of refracted shock waves within the chamber and the critical role of charge ignition geometry in their causes and prevention. This empirical understanding is clear in Luis Collado's discussion of the adverse consequences of a touchhole placed too far forward (Collado, 1592: 10). Nineteenth century measurements of peak pressures give us our best idea of the magnitude and destructive capabilities of the forces involved, and one of the major achievements of black powder ballistics in its final stages was the success achieved in controlling them.

Medieval and early modern gunners believed, correctly, that large-grained powder was preferable for use in large ordnance, but the advantages of this preference were lost with the general adoption of eprouvette mortars and short-barrelled powder testers in the 18th century. This led to a 'rational' appreciation that fine-grained musket powder was more powerful than the old, 'grosse',

cannon powders — as indeed it was in very short guns — and they were abandoned. The reasons are worth touching on. Since eprouvette mortars and powder testers had very short barrels, they measured the pressure developed by the charge in the first few milliseconds of burning rather than the amount of propulsive energy in the powder. While they were no doubt an effective means of detecting deterioration in a given batch of powder, they militated against large-grained, slow burning powders. It was not until the late 1850s that Rodman showed the fallacy involved, pointing out *the impropriety of taking the eprouvette range as an indication of the projectile force of powder which is to be used in guns of any considerable length and calibre* (Rodman, 1861: 204).

As the medieval gunner suspected, and Rodman proved experimentally in the late 1850s (Noble, 1871: 8), fine-grained, fast burning powder places excessive strains on heavy ordnance. Since the propulsive force applied to the ball varies with the square of the bore and the mass resisting it varies with the cube of the ball diameter, projectile acceleration for a given pressure is an inverse function of bore diameter, the larger the bore the less the acceleration for a given chamber pressure.[17] A charge of fine-grained powder in a large gun converts its chemical energy into gas before significant projectile movement can occur and the pressure peaks before the ball can move to relieve it by expanding the volume. The bigger the gun the more acute the problem: large charges are thermodynamically more efficient than small ones, so more pressure is produced for a given weight of powder. In addition, more powerful shock waves can develop in a large chamber, a problem which Rodman's results demonstrate, though he was cautious in attributing cause.[18] If, however, the grain size is sufficiently large, the projectile will begin to move while the chamber pressure is still rising, enlarging the volume within which the gas is evolving and thus reducing the peak pressure.

Large-grained powders thus place less strain on the gun for the amount of propulsive energy delivered to the projectile, expressed in practical terms as muzzle velocity. Rodman summarized the results of the experiments which led him to this conclusion in these terms:

> ... *the maximum pressure of gas diminishes as the diameter of grain increases, in a much greater ratio than the squares of the corresponding velocities; this showing, conclusively, that the velocities due to our present charges of small grained powder may be obtained with a greatly diminished strain upon the gun, by the use of powder properly adapted in size of grain to the calibre and length of bore in which it is to be used* ... (Rodman, 1861: 204).

Rodman went so far as to test powder pressed into thick cylindrical plates which completely filled the bore, with a small hole through the center for ignition (H.M. War Office, 1895: 21–2). This powder worked well, but the large grains proved too brittle for field service and the Americans turned to powder pressed into very large spheroids for large ordnance, the so-called walnut grain powder. The later British prismatic powders were pressed into perforated hexagonal plates which could be stacked tightly in a bagged charge, approximating Rodman's perforated cake in performance.

In sum, the ballistic performance of black powder depended upon a considerable number of complex physical, chemical and thermodynamic variables, the operation of many of which is beyond our power to explain. For a sense of their interaction, we must turn to the 19th century ballisticians who were the last to experiment directly with them in large ordnance. It would be difficult to improve on the wording of the 8th edition of the British War Office's *Treatise on Ammunition* (H.M. War Office, 1905):

> *Speaking generally, a large grained powder, highly glazed, made from highly-burnt charcoal, and of high density, other things being equal, will burn more slowly, and therefore be less violent in its action than a powder of opposite characteristics.*

The same passage continues with a characterization of the performance of very finely-grained powders, providing a useful starting point for speculation on the performance of serpentine:

> *If, however, the grain be very small as in mealed powder, the interstices between the grains are not sufficiently large to allow a free passage of flame, and so a charge of mealed powder would ignite in one place only and would burn comparatively slowly.*

A change of serpentine in a large gun, ignited at the rear, would no doubt disperse as it travelled down the barrel, but whether the burning rate would then fall off, accelerate catastrophically, or remain constant must remain a matter of speculation, barring experimentation.

Efforts to apply experimental science to internal ballistics began in the 18th century. The French experimenter du Mé in 1702 described a test in which he exploded a measured quantity of gunpowder in the sealed, upper end of a V-shaped container filled with water to permit measurement of the quantity of evolved gasses by the upward displacement of water in the other, open end. He found that the gasses evolved by a cubic inch of gunpowder displaced 4,000 cubic inches of water (Muller, 1973: 212). Benjamin Robins continued this line of experimentation in the mid-18th century, and determined that gunpowder evolved the same amount of gas in air or in a vacuum (du Mé had believed the expansion to be due largely to heat imparted to the air between the particles of gunpowder). Robins estimated that gunpowder would produce about 240 times its volume of gas and, if contained in its original volume, would yield a pressure of 1,000 atmospheres (14,700 psi)

(Muller, 1973: 212). In the event, Robins' figures were far off the mark (Nobel, 1871: 3–4) but this was of little practical consequence for the means to apply such data to practical engineering problems were lacking and would remain so for a long time to come.

Robins' greatest contribution was the ballistic pendulum, noted above in connection with external ballistics. It was only large enough to test musket balls, and deformation of the ball and the changing angle of the plate as the ball rebounded were no doubt sources of inaccuracy, but the principles on which it was based were sound and the concept was systematically developed (Douglas, 1855: 28). By the 1850s the velocities of cannon projectiles were measured by firing them into pendulums of massive suspended baulks of timber. The projectile lodged in the baulk, converting its kinetic energy into potential energy, expressed as upward movement against the force of gravity; the height of the pendulum's swing could be easily measured and indicated the amount of potential energy. From that point, computing the impact velocity of the projectile was straightforward. Ballistic pendulums became remarkably sophisticated: the great mass of the baulk reduced vibration and kept velocities low, minimizing the effects of aerodynamic drag; the height of swing was traced virtually without friction in a vertical bed of grease by a stylus mounted on the baulk. One inherent problem with the ballistic pendulum was that the ball had to travel a significant distance under the influence of aerodynamic drag before impact if the pendulum were not to be affected by the propellant gasses expelled from the muzzle behind the projectile, and ahead of it as well in the case of smooth bore ordnance. Mid-nineteenth century experimenters recognized that aerodynamic drag was greatest at the highest, initial velocities and, lacking analytical methods to accurately predict drag, expended considerable ingenuity in attempts to minimize its experimental effect. Their methods included mounting cannon and pendulum close together and firing the ball through a thin sheet of lead. Cannon themselves were turned into ballistic pendulums, suspended so that the axis of the bore remained parallel to the earth during recoil (Douglas, 1855: 29–49, Plate 1, 49–50). From the energy computations outlined above and conservation of momentum, this permitted estimation of the total propulsive energy developed by the gun as well as the theoretical maximum projectile velocity. Comparing these values with the impact energy of the projectile, taken with a conventional ballistic pendulum, permitted computation of the net efficiency of the gun and estimations of the amount of energy lost to escaping gas.

A detailed analytical chronology of the developments in question has not been constructed, but it is clear that the ballistic pendulum first affected the practical art of gunnery by permitting direct measurement of the velocity imparted to the ball by powder charges of varying sizes and compositions. The optimum size of powder charges was soon found to be considerably smaller than prevailing practice and, at some point along the way, the effect of the quality of charcoal on gunpowder's propulsive energy was measured directly.

Charges were reduced in size, with immediate benefits in safety and logistics, but this merely verified and accelerated a process which had been proceeding on empirical grounds since the 16th century (Lewis, 1961: 201). Of more importance, the ballistic pendulum gave unequivocal evidence for the greater propulsive powers of gunpowder made with charcoal reduced in enclosed iron vessels, rather than burned in open ricks or pits as had been the case previously. This discovery marks the birth of experimental ballistics as a science with practical engineering application. The reasons for the greater efficiency of powder made with cylinder charcoal, as it was called, might not have been understood, but its superiority could not be doubted. The ballistic pendulum did not lie. In Britain, where the discovery was made, the efficiency of heavy ordnance was a matter of great concern to the state and the message was not missed. The adoption of cylinder charcoal was first suggested in 1783 and powder made with it, called cylinder powder, was introduced into the Navy about 1800 (Lavery, 1987, 135).

The next major watershed in internal ballistics came in the late 1850s with the development by Rodman of a means of measuring pressures inside the bore (Noble, 1871: 8). Rodman's method was straightforward and ingenious: he drilled cannon at right angles to the bore, tapped the holes, and mounted piston assemblies in them which trapped internal pressure and harnessed it to drive a knife-edged indenting tool against a copper specimen (Figure 1). Since the amount of pressure required to create an indentation of a given depth in a standard specimen was known, this permitted direct reading of peak pressures (Rodman, 1861: 174–5). This was the first means of directly measuring events within a gun (Figure 2).

Rodman's tool was later found to be sensitive to transient pressures, in effect permitting the pressure to impart kinetic energy to the piston assembly before the indentation was completed. It was eventually replaced by the crusher gauge (Figure 3), in which the knife edge and specimen were replaced by a small copper cylinder which was crushed by a flat-headed piston against a parallel surface (Noble, 1871: 19). The peak pressure was calculated from the amount of compression, or crush, which the copper cylinder sustained along the longitudinal axis. The crusher gauge proved to be more accurate than Rodman's tool, but the principles behind it were the same and later ballisticians acknowledged their debt to Rodman (Figure 4) (Noble, 1871: 8).

The first significant fruit of Rodman's experimental method was his unequivocal demonstration of the advantages of large-grained powder in large ordnance and since his methods and his results were intimately bound up with one another, the latter bear repeating in part. These are given in Table 2

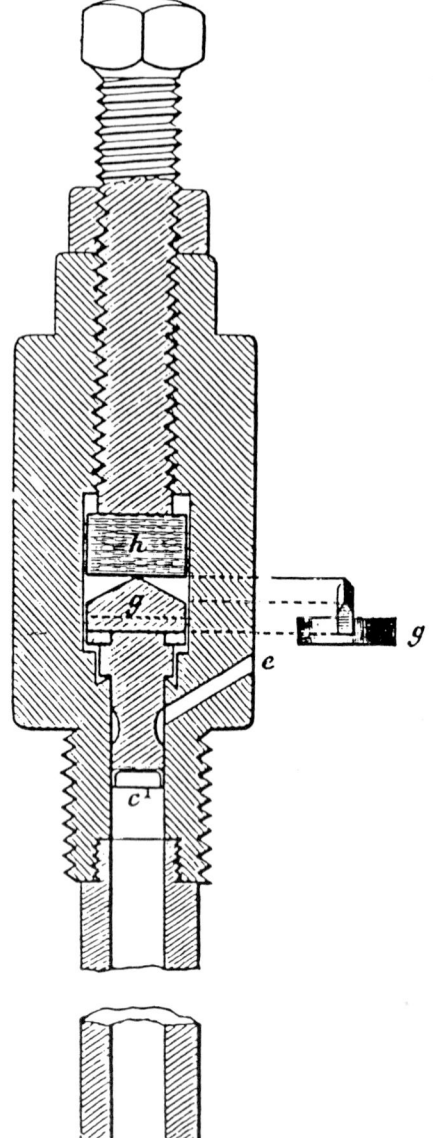

Figure 1   Rodman gauge.

as they were later presented by Andrew Noble, Rodman having published his results in tabular form. The test was performed with ordinary, operational 42 pounder smooth bore cannon and Rodman recorded the muzzle velocities. His findings in this experiment thus not only represent a major benchmark in the development of ballistics as an applied science, but have historical interest in their own right,

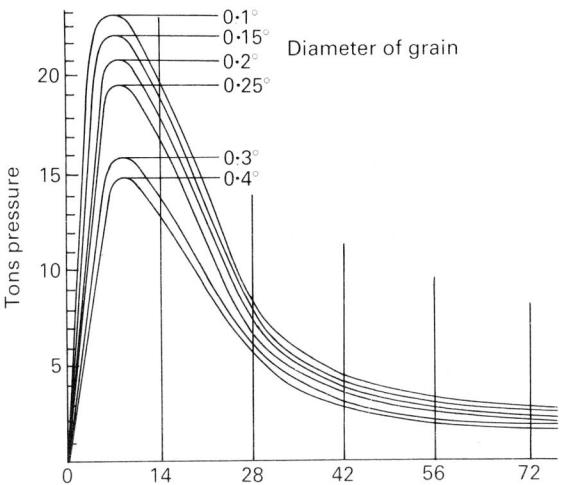

Figure 2   Pressure as a Function of Distance from Base of Chamber.

giving us a sense of the capabilities of black powder service ordnance of a size with considerable historical importance.

The pressures, expressed in tons of 2,240 lbs/in^2, are given in increments of distance rather than time since Rodman could only measure peak pressure at given points along the bore. Nevertheless the results yield the same essential information as a time/pressure plot, and, while they could not be used to calculate the net propulsive energy developed within the gun as Noble was to do with time/pressure curves, they gave an accurate qualitative representation of the process which Rodman was to use to good effect. It only remains to note that the ordinary service powder, with a 0.1 in grain diameter, was the finest of the six tested. The results, considered in conjunction with the muzzle velocities, given below, speak for themselves. A small additional quantity of large-grained powder would bring the muzzle velocity up to the value obtained with the fine-grained powder, and would do so with considerably less stress to the gun.

Rodman also investigated the effect of windage on ballistic performance, a matter of enduring concern to practical gunners and

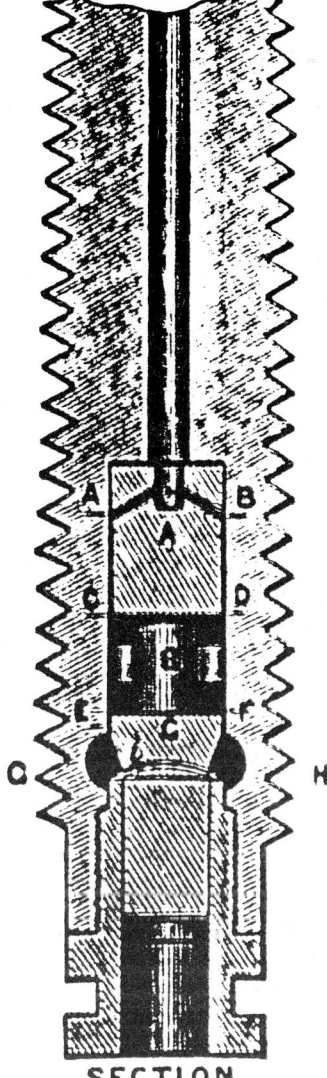

SECTION
Figure 3   Crusher gauge.

cannon founders alike, perhaps because it was so easily measured. Conventional wisdom was that increased windage would reduce stress on the gun by relieving internal pressures, and in tests with a 7 in (42 pounder) cannon Rodman found that increasing the windage from the normal 0.18 in to 0.3 and 0.4 in yielded a progressive decline of 3.2% in chamber pressure and 3.5% in muzzle velocity (Rodman, 1861: 109). The results are qualified by the fact that the windage was increased by grinding one

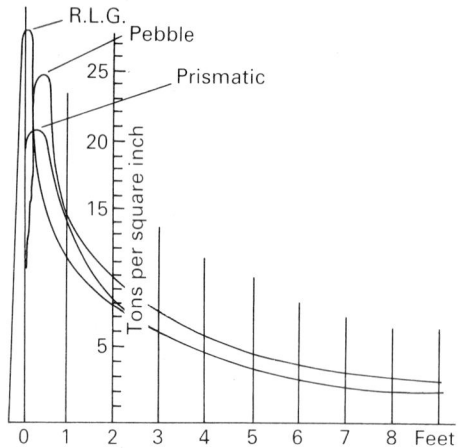
Figure 4   Pressures as a Function of Time and Distance.

side of the ball flat rather than by using smaller balls and the maximum pressures recorded may have been transient peaks. In sum, the results were equivocal and, as with so many other black powder performance issues, we are unable to make detailed predictions concerning the effect of variations in windage on ballistic performance.

Rodman did not confine his activities to internal ballistics; indeed, he was probably better known at the time for pioneering the technique of casting iron cannon around a chilled core. The internal stresses set up by differential cooling substantially increased the strength of the guns and permitted the casting of larger ordnance than previously. This takes us beyond our focus on ballistics but for the fact that Rodman incorporated his experimental findings concerning the variation of press-

ure as a function of distance down the bore into the design of large ordnance. Rodman's large guns were proportionately shorter than had been the general rule for heavy ordnance previously; they had no muzzle flare and the thickness of the barrel was proportionate to the anticipated peak pressure at each point. This gave them a rounded, tapered look not unlike the guns of the *Milamete* manuscript.

During the American Civil War, leadership in experimental ballistics shifted to Europe, where it remained until the end of the black powder era. Experimental methods continued to develop, particularly in Britain, where black powder internal ballistics probably attained its highest development toward the end of the century. The most significant advance in experimental method during this period was the application of electricity to measure projectile movement within the bore, and here Andrew Noble took the lead, adapting the Wheatstone chronograph, invented about 1840, to internal ballistics ('Chronograph', 1911: 302).

The requisite measurements were taken by drilling into the bore, as with Rodman's pressure gauge, and inserting plugs which contained a fine electric wire which was severed mechanically by the passage of the shot. The wire in question was part of the primary of a battery-powered induction circuit, the secondary of which was arranged so that when the primary was severed, a spark would arc between two ends of the interrupted secondary. These were mounted in close

Table 2
Mean Muzzle Velocities Obtained With Powder of Various Grain Diameters

	Muzzle Velocities					
Grain size	1 in	1.5 in	2 in	2.5 in	3 in	4 in
Mean velocity	1261 fps	1235 fps	1199 fps	1151 fps	1146 fps	1187 fps

Velocities were obtained with a ballistic pendulum. Five firings were undertaken with each grain size and a minimum of three valid velocity reading obtained.
(Rodman, 1861: 199–201).

proximity to one another, separated by a whirling paper disk, in which the spark left a minute hole (Noble, 1871: 19–23). The disk was one of eight mounted on a common shaft driven at a high and carefully regulated speed by a falling weight. Each disk was connected to an induction circuit as described above, and the plugs were inserted into the bore at intervals.

The details of this remarkably sophisticated apparatus need not concern us further; the essence of the thing was that the shot severed the primary circuits in sequence in its passage down the bore and the time lapse between them could be determined by measuring and comparing the locations of the holes in the disks. The disks moved at a speed of some 1,200 in/sec (30.5 m/sec) at the point of arcing, and in principle time intervals of less than a millionth of a second could be measured (Noble, 1871: 20). In practice, the device was accurate to a hundred thousandth of a second, and this was more than adequate ('Chronograph', 1911: 303). Knowing when the shot had passed each plug, Noble and his collaborators could construct a time/velocity curve for the movement of the projectile within the bore with useful accuracy. From velocity and time, they could calculate the acceleration of the projectile at any given point, and from acceleration compute the propulsive force. Since the propulsive force was a function of pressure, as discussed above, it was now possible to draw a continuous plot of net effective pressure at the base of the projectile for its entire passage down the bore.

The energy required to impart a given velocity to a projectile is a function of the area beneath the time/pressure curve (c.f. Greenhill, 1911: 276–77, based largely on Noble's work). That curve could now be determined experimentally and the results checked by measuring the muzzle velocity, accomplished electrically with considerable accuracy by having the projectile cut a series of wires as it exited the muzzle. Since the muzzle energy of the projectile could be accurately calculated, the effective propulsive energy could be compared with that indicated by the area under the time/pressure curve and sources of inefficiency such as heat loss to the barrel, friction, and so forth, closely estimated. More important, once these factors were taken into account, net pressures could be calculated with considerable accuracy and compared to crusher gauge pressure readings. This permitted more accurate calibration of cruser gauge readings, further enhancing experimental accuracy in synergistic fashion.[19] Of more immediate concern, the existence of shock-wave induced pressure peaks was unequivocally confirmed and their surprising amplitude and extreme variability, previously masked by instrumental error, made clear. Also apparent was the clear connection between pressure 'spikes' and fine-grained, fast-burning powders (Noble, 1871: 31–33). The discovery was timely as transient pressure peaks posed an increasingly serious design problem as naval guns, in particular, became much larger and more powerful with the accelerating race between armour and penetrating power from the 1870s on.

The last smooth bore cannon in British service had muzzle velocities in the order of 1,600 fps (490 m/s). These dropped to some 1,200 fps (370 m/sec) with the adoption of rifling and elongated projectiles, but crept steadily upward to about 1,400 fps (430 m/sec) where they stagnated for a time until governments intervened in the 1850s, notably through Rodman's work in the United States (though Mordecai, Dahlgren and others made significant contributions), and through the establishment of a committee on explosives, of which Andrew Noble was a member, in Britain. In large part through the government-supported efforts of these individuals and bodies, velocities were progressively raised to around 1,600 fps (490 m/sec) by the 1870s. While these velocities are not particularly impressive by today's standards, the increase represented a rise in muzzle energy of about a third and was achieved without any revolutionary breakthroughs in structural materials. Impressively, the increase in muzzle velocities was accompanied by a significant reduction in maximum chamber pressures of about the same proportions, that is a reduction of about a third in the British case (Noble, 1907: 4).

After the American Civil war, interest in ballistic research waned in the United States. From that point, Britain, if not always ahead in every category, increasingly set the pace in the development of ordnance through to the end of the black powder era. The pace of technological change accelerated, the role of innovative and well-capitalized private entrepreneurs became increasingly important and changes in materials and design practice began to feed on one another interactively. This was nowhere more true than in the design and production of heavy ordnance, particularly naval ordnance. By 1890 velocities in the order of 2,000–2,100 fps (610–640 m/sec) were common for armour-piercing guns (Noble, 1890: 11). Such velocities, as I suggested earlier, were probably close to the absolute limit attainable with black powder. Even more impressive, charges above 1,000 lbs (454 kg) were routinely used (Noble: 1890: 11). The potential for catastrophe inherent in the immense energies which these figures imply was great and the final developments in black powder manufacture were clearly driven by the desire to reduce the magnitude of transient peak pressures or prevent them altogether. The development and adoption in British service of prismatic cocoa powder, made of 79% KNO_3, 19% overburned charcoal and only 2% sulphur is evidence of this (Noble, 1907: 1, 6). Indeed, in a sense the development of such sophisticated propellants for large ordnance might be considered the apex of black powder internal ballistic development.

This brings us, full circle, to considerations of the composition of gunpowder and its effect on the shape, size and performance of guns. In essence, the internal ballistic properties of black powder combined with the materials available for the construction of barrels to dictate the physical and operational characteristics of artillery. These parameters, in particular, determined muzzle velocity and hence impact energy. External ballistics combined with projectile characteristics and carriage design to determine accuracy. These factors determined the net tactical characteristics of artillery in the black powder era, and they remained remarkably constant across its entire span. I have only sketched the high points of the development of black powder ballistics and have knowingly left much out . . . and no doubt unknowingly as well. My fundamental concern throughout was to relate the problems of the design, construction and use of ordnance, particularly heavy ordnance, to the characteristics, capabilities and behaviour of the material which shaped the actions of gunner and engineer, both creating and limiting the possibilities open to them: black powder.

The brief flowering of black powder internal ballistics coincided with the birth of applied science and, indeed, may have precipitated it. The developments which I have reviewed in cursory fashion should be considered within this context. Plainly, the preemptive priorities of war have exercised a strong influence on the development of technology from the earliest times, and gunpowder occupies a position of particular importance and interest in assessing man's efforts to develop and harness new and more potent sources of energy. Experimental science assumed a significant role in the process relatively late and the pivotal development in the process was an invention of elegant simplicity, Benjamin Robins' ballistic pendulum.

Ironically, while Robin's main interest seems to have been in external ballistics — he became a champion of the so-called 'short gun' school of naval ordnance, largely on the basis of his appreciation of external ballistic considerations (Douglas, 1855: 102) — his ideas had their most significant technological impact in internal ballistics. It was here that ballistics developed as an applied science and while Rodman's pressure gauge and Noble's ballistic chronograph may fall behind the ballistic pendulum in intellectual significance, they may still be seen as remarkable early successes in the systematic application of scientific theory to the solution of engineering problems. Indeed, in the transition from Robin's first, crude, ballistic pendulum to Noble's painstakingly researched and carefully worked out velocity curves and pressure computations, we can observe the birth of engineering as an applied science.

Notes

1 European gunpowder recipes specified ingredients by weight from the very beginning. However early recipes are not precisely comparable since the purity of the sulphur and, particularly, the saltpetre is unknown. Tartaglia, writing around 1540 believed that the very earliest gunpowder was mixed to a ratio of 1/1/1 and Partington speculates that this may have been so based on his analysis of Roger Bacon's late 13th century recipe which calls for proportions of 7/5/5 (Tartaglia, 1588: 72; Partington, 1960: 74–8). I organized and observed a series of black powder test firings at the H. P. White Laboratory, Bel Air, Maryland, on 1 July 1969, including a limited number of firings with powder prepared by me according to early recipes. A 552 grain charge of 1/1/1 serpentine black powder, fired behind a 0.812 inch diameter steel ball in a 0.86 inch diameter smoothbore barrel 31 inches long, produced a muzzle velocity of 117 fps. This fizzle is meaningless in isolation, however a similar test with an equal charge of homemade 75/15/10 powder produced a velocity of only 112 fps. Only light paper wadding was used in these firings and two subsequent tests with charges of the same homemade 75/15/10 powder, tightly confined by driven-in wooden plugs, produced muzzle velocities of 1350 fps and 1340 fps respectively. By comparison, two firings with 552 gr of modern fffg sporting powder produced an average velocity of 1935 fps and seven firings with 276 gr of fffg and the same ball/wadding combination averaged 1638 fps. A tightly confined charge of 276 gr of homemade $50/33\frac{1}{3}/16\frac{2}{3}$ powder produced a 118 fps fizzle. Experiments by Dr A. R. Williams, with serpentine and wet-incorporated powder mixed to a ratio of 6/2/1 in a 0.75 in (19 mm) barrel, showed significantly higher average muzzle velocities with the wet-incorporated powder (930 fps as opposed to 870 fps, firing a steel ball bearing from a 15 in (38 cm) barrel); the peak velocities obtained with the two powders, however, were remarkably high and surprisingly close, over 2,300 fps (700 m/sec). Significantly, these very high velocities were all attained with a firmly driven-in lead ball (Williams, 1974).

2 The argument for the transfer of gunpowder technology from China to Europe is strengthened by the similar appearance of the very earliest Chinese and European depictions of gunpowder weapons. The Chinese evidence, from statuary dating as early as A.D. 1128, shows a stubby, flask-like weapon strikingly similar in shape to the two guns depicted in the Walter de Milamete manuscript of 1327 (Gwen-Djen, Needham and Chi-Hsing, 1988: 596, 601–2). It has been suggested that the Milamete date is that of the manuscript, the illustrations proper being undated (Cipolla, 1965: 21). To support the idea that manuscript and illustrations are contemporary, the armored figure firing the smaller weapon wears ailettes, flat plates lashed to the shoulder points of the mail hauberk. Ailettes appeared around the middle of the 13th century, were characteristic of armor of the 1320s, were going out of fashion by the beginning of the Hundred Years War and had almost disappeared by the 1340s (Norman and Don Pottinger, 1985: 65, 87–88; Oman, 2: 377 n. 2). Assuming that the humans depicted with them are to scale, the larger of the Milamete weapons was some 8 feet ($2\frac{1}{4}$ meters) long and about a third that large in diameter while the smaller was $3\frac{1}{2}$–4 feet (1–$1\frac{1}{4}$ meters) long and half that large across, or perhaps a bit more. The Chinese gun, held by a human-like demon, is about the same size as the smaller of the Milamete pieces, though the statuary is allegorical. The European projectiles are over-sized crossbow bolts which, we can adduce from other sources, were almost surely wrapped with a leather strip to provide a gas seal with the bore, while the Chinese projectile is apparently spherical. The Milamete illustrations contain no suggestion that the guns were bound or otherwise fixed to their 'mounts', insubstantial-looking trestle tables. The impression throughout is of limited recoil and propulsive capacity.

3 I have used the term nitrocellulose-based as shorthand; nitrocellulose propellants proper, made by nitrating cellulose compounds such as cotton, were followed by similarly-made propellants based on other organic substances. These included nitroguanidine, originally based on guano, and the so-called double base propellants incorporating nitroglycerine. Chemically and ballistically, their behavior is broadly similar and differs sharply from that of black powder.

4 The argument that there were tactically significant range differences among cannon used in the Armada campaign and that the range differences were attributable to relative barrel length was advanced by Michael Lewis in a series of articles in *Mariner's Mirror* in 1942–43, published in 1961 as *Armada Guns* (Lewis, 1961: esp. 189, n. 3). I questioned this assumption in 1974 in *Gunpowder and Galleys*. (Guilmartin, 1982: 140; Guilmartin, 1988: 47–51). My arguments did not go without challenge (Barker, 1983: 68–70), and recent books on the Armada campaign reflect a lack of consensus, ranging from uncritical acceptance of Lewis' assumptions (Fernandez-Armesto, 1988: 61–2) through recognition of the debate as unresolved (Padfield, 1988: 85–88) to explicit acceptance of my position (Martin and Parker, 1988: 196–7).

5 Noble records a velocity of 2497 fps for a 30 lb shot fired in a 5.87 inch, 32 calibre barrel with a charge of relatively fine-grained R.L.G. powder (Noble, 1907: 8), the highest value I have encountered. It was recorded as part of a test series designed to probe extreme capabilities and was not representative of service weapons. The upper limit of muzzle velocities for ordinary operational weapons was probably about 2,000 fps.

6 Artillery design was approaching an empirical optimum for the materials available, and ordnance cast in a royal foundry for a given warship may reasonably be considered to have been cast to the same specifications. It is thus worth noting that the bores of lighter, upper deck, pieces were markedly longer in calibres than those of the heavier lower deck pieces (e.g. Guilmartin, 1983: Fig. 4a; Boudriot, 1986: 158). Partly, of course, the relative shortness of the heavier guns is no doubt attributable to attempts to reduce weight and make the guns easier to handle. But it can also be taken as empirical evidence that optimum barrel length is an absolute, rather than a relative value.

7 Smaller barrels absorb a greater proportion of propulsive energy as heat, but the net effect on ballistic performance of black powder ordnance has not been explored in detail. Rodman's generalization in this regard is probably correct, but his pressure measurements in the tests in question were probably transient peaks; see note 18, below, for a fuller discussion.

8 The relevant concept is specific impulse, I_{sp}, a standard measurement of rocket propellant efficiency. Calculated in (lb/)lb/sec) or the metric equivalent), I_{sp} expresses the number of pounds of thrust produced per pound of fuel per second; the units, seconds, indicate efficiency not time, the higher this value the more efficient the fuel (Huzel and Huang, 1971: 10–11). From $F = ma$, fuels with less massive decomposition products yield higher specific impulses, or, in plain language, more acceleration per unit weight. As a rocket fuel, black powder yields an I_{sp} of 50–70 sec; double base solid propellants similar in composition to smokeless powders yield values of 180–210 sec and common liquid rocket fuel combinations yield values of 230–270 sec (Durant, 1973: 409, 422). Since the decomposition products in a rocket exhaust are in effect projectiles, exhaust velocities are limited almost entirely by the molecular weights of the decomposition products and rocket exhaust velocities provide a crude comparative measure of maximum muzzle velocities theoretically attainable in guns using the same propellants. Expected black powder exhaust velocities lie in the range of 2,900–3,250 fps (900–1000 m/sec) while those of smokeless powders are about 4,900–5,900 fps (1,500–1,800 m/sec) (Ley, 1945: 213).

9 I am unaware of any direct evidence concerning the density of medieval or 16th century powder charges, however Collado's pattern for a full charge cartridge for a culverin has the same diameter as the ball and is four ball diameters long (Collado, 1592: 47). Being of slightly less diameter than the bore, it would be compacted by ramming. Douglas gives a density of 58 lbs/ ft³ for artillery powder in the period before the adoption of large-grained, compressed powders (Douglas, 1855: 475) and a charge of this density equal to

the weight of the ball would occupy just under $2\frac{1}{4}$ calibres.

10 The Catalan forge, generally held to have appeared around 1300, represented the first systematic use in Europe of convective draught through a vertically stacked charge of ore and fuel to produce wrought iron directly in a single step (Elliott, 1987: 363). Until the advent of the Catalan forge, Chinese iron-working techniques were probably ahead of those in Europe and the earliest Chinese bombards may have been made of cast iron (Gwei-Djen, Needham and Chi-Hsing, 1988: 604). There can be little doubt that the Catalan forge had a significant positive effect on early firearms design and manufacture in Europe. The key role of charcoal in both gunpowder and the Catalan process is suggestive.

11 Modern nickle bronzes have melting points in excess of $1220°C$ ($2230°F$), but the melting points of classical and medieval bronzes were probably below that of pure copper, $1083°C$ ($1981°F$) (Kent, 1938: 4–29). The figures for iron are for a modern cupola furnace with draught provided by blowers (Kent, 1938: 20–10, 20–13), but the figures are probably within the range of those produced by a medieval Catalan furnace.

12 The value in question was obtained with relatively fast-burning powder and a 360 lb shot in a ten inch rifled gun. In the same series a 120 lb shot, roughly equivalent in weight to a spherical cast iron ball for the same bore and thus more indicative of the stresses to which earlier, smoothbore cannon were subjected, yielded a chamber pressure of 2,830 atmospheres. These were averages, obtained by calculating the force required to produce the observed acceleration of the projectile within the bore. Noble routinely recorded peak chamber pressures of 4,200 atmospheres (61,600 psi; $4.33 \times 10^7 \text{ kg/m}^2$) by direct measurement in the same series (Noble, 1871: 39) and noted transient peak values as high as 9,500 atmospheres (139,500 psi; $9.19 \times 10^7 \text{ kg/m}^2$) (Noble, 1907: 4).

13 The prime culprit, though by no means the only one, was the sharp, non-linear variation of the drag coefficient at velocities near the speed of sound (e.g., Guilmartin, 1974: 278, Fig. 15). Since most artillery and small arms projectiles exited the muzzle at speeds in excess of sonic velocity (1096 fps or 334 m/sec at sea level on a standard day) this problem could not be avoided.

14 Douglas' summation of his disagreement with Benjamin Robins' advocacy of shorter naval guns with lower muzzle velocities, argued on the basis of sound ballistic principles, is of particular interest and underlines the extent to which practical considerations dominated (Douglas, 1855: 102–3). One of Douglas' most telling arguments against carronades is their proclivity to set rigging and hammocks afire with muzzle blast, a product of their short barrels.

15 Until 1978, if not since, the U.S. military specification for black powder required that it be made of charcoal burned from willow or alder and imposed no further requirements for chemical composition (Harris, Lannon et al, 1978: 355).

16 Dr Williams encountered persistent problems with slow and erratic ignition in his tests with serpentine; these encompassed some 28 shots, enough to support generalization. He noted that tight initial confinement

appeared to promote faster and more complete ig-nition and that on several occasions most of the force of the charge escaped out the $\frac{1}{8}$ in (32 mm) touchhole, relatively large in relation to the bore.

17 The driving force, F, is proportional to the pressure behind the ball, p, times the area of the bore, $F = 1/4\,\pi\,d^2 p$. The weight of a ball of diameter d made of a material of density ρ is given by $w = 1/6\,\pi\,d^3\,\rho$ and the mass equals the weight divided by the gravitational constant, g, 32.2 ft/sec/sec. We can consider ball and bore diameter to be the same and ignore the effects of windage, which are small in any event.

From $F = ma$

$$\left(\frac{\pi d^2}{4}\right)p = \left(\frac{\pi d^3\,\rho}{6g}\right)a$$

which reduces to

$$a = \left(\frac{3g}{2}\right)\frac{p}{d\rho}$$

or in English units, $a = 48.3\dfrac{p}{dp}$

In other words, for a given pressure the acceleration imparted to the ball varies inversely as a function of the diameter of the ball and its density. This set of relationships, while simplified, approximates reality and effectively illustrates the basic internal ballistic problem posed by increasing the size of ordnance. It also helps to explain the early preference for stone projectiles in large ordnance.

18 Rodman conducted a test in 1859 with a 7 inch, a 9 inch and an 11 inch gun with the same chamber geometry and bore lengths, firing the same weight of powder behind the same weight of shot per unit of bore area. His peak chamber pressure readings aver-aged 58% higher in the 11 inch gun and 45% higher in the 9 inch gun than for the 7 inch gun (Rodman, 1861: 184, 195–9). For reasons we will address below, Rodman's highest pressure readings were surely the result of transient peaks caused by refracted shock waves. This hypothesis is supported by the fact that there was no significant difference in average muzzle velocity among the three guns, by the fact that devi-ations from the average maximum chamber pressure were markedly greater in the two larger pieces and, most conclusively, by the fact that there was no corre-lation whatever between peak chamber pressure and muzzle velocity, notably for the shots on which the highest and lowest peak pressures were recorded for each gun (Rodman, 1861: 184, 197). Rodman's data on this test series is exhaustive (pressure readings were taken at seven points along the bore for each shot) and bears further scrutiny in light of current ballistic theory and methods of statistical analysis. Noble's analysis of Rodman's methods and critique of his data, prompted in large part by the results just outlined, is still thought-provoking (Noble, 1981: 8–12). Rodman's integrity in publishing admittedly anoma-lous data and Noble's graciousness in acknowledging his intellectual debt to Rodman speaks well for both men.

19 These were probably on the whole low, since copper deforms less under a sudden transient force than an equivalent static pressure, and with smokeless propel-lants in the 1960s crusher gauge readings were found to represent as little as 80% of the actual pressure sustained (U.S. Army, 1964: 67).

References

Barker, R, 1983, 'Bronze cannon founders: comments upon Guilmartin 1974, 1982, *IJNA*, **12**.1: 67–74.

Benton, J G, 1862, *A course of instruction in ordnance and gunnery complied for the use of the cadets at the U.S.M.A.* New York.

Biringuccio, V, 1942 (based on the 1540, Venice, edition), *The pirotechnia*, C S Smith and M T Gnudi, tr. New York.

Blackwood, J D and Bowden F P, 1952, "The initiation, burning and thermal decomposition of gunpowder', *Proceedings of the Royal Society*, series A, *Mathematical and Physical Sciences*, **CCXII**. 1114.

Boudriot, J, 1986, *The seventy-four gun ship*, vol. 2. D Roberts trans. Paris

"Chronograph", 1911, *Encyclopaedia Britannica*, 11th ed., **V**: 422. New York.

Cipolla, C, 1965, *Guns, sails and empires: technological innovation and the early phases of European expansion 1400–1700.* New York.

Collado, L, 1592, *Platica manual de artilleria.* Milan.

Douglas, H, 1855, *A treatise on naval gunnery, 1855* (Conway Maritime Press 1982 reprint). London.

Durant, F. C. D, III, 1973, 'Rockets and guided missiles,' *Encyclopaedia Britannica*, **19**: 403–24.

Elliott, J F, 1987, 'Iron production', *Encyclopaedia Britannica*, 15th ed., **21**, 360–88.

Eschbach, O W, 1966, *Handbook of engineering fundamen-tals*, 2nd ed. New York.

Fernandez-Armesto, F, 1988, *The Spanish Armada: The experience of war in 1588*, Oxford.

Greenhill, A G, 1911, 'Ballistics', *Encyclopaedia Britannica*, 11th ed., **III**: 270–79. New York.

Guilmartin, J F Jr, 1988, 'The internal ballistics of black powder', appendix to 'Early modern naval ordnance and European penetration of the Caribbean: the oper-ational dimension', *The International Journal of Nautical Archaeology*, **17**.1: 47–53.

Guilmartin, J F Jr, 1983, 'The guns of the *Santissimo Sacramento*', *Technology & Culture*, **24**.4: 559–601.

Guilmartin, J F Jr, 1974, *Gunpowder and galleys: changing technology and Mediterranean warfare at sea in the sixteenth century*, Cambridge.

Gwen-Djen, L, Needham J and Chi-Hsing P, 1988, 'The Oldest Representation of a Bombard,' *Technology and Culture*, **29**.3: 594–605.

Hall, A R, 1983, 'Gunnery, science and the Royal Institution', *The uses of Science in the age of Newton*, J G Burke, ed. Berkeley.

Harris L E, Lannon, J A, Field R and Husted D, 1978, 'Spectroscopic investigation of the combustion of black powder', *Journal of Ballistics*, **2**: 353–91.

Held, R, 1957, *The Age of Firearms*. New York.

H.M. War Office, 1895, *Treatise on service explosives*. London.

H.M. War Office, 1905. *Treatise on ammunition*, 8th ed. London.

Huzel, D K and Huang, D H, 1971, *Design of Liquid Propellant Rocket Engines* (NASA SP–125), 2nd ed. Washington, D.C.

Kent, R T, 1038, *Kent's mechanical engineers' handbook, design. Shop practice*. 11th ed. New York.

Lavery, B, 1987, *The arming and fitting of English ships of war 1650–1815*. London.

Lewis, M, 1961, *Armada guns, a comparative study of English and Spanish armaments*. London.

Ley, W, 'Evaluating the vaunted V2' *Aviation*, vol. 44, no. 2, February 1944, pp. 212–214.

Martin, C and Parker, G, 1988, *The Spanish Armada*. London.

McNeill, W H, 1982, *The pursuit of power, technology, armed force, and society since A.D. 1000*. Chicago.

Muller, H G, 1973, 'A brief history of powder testers', *Arms and armor annual*, Robert Held, ed., **1**: 206–215. Northfield, Illinois.

Noble, A, 1871, *On the tension of fired gunpowder, being a lecture delivered at the Royal Institution, March 3, 1871*. London

Noble, A and Abel, F A, 1874, 'Researches on explosives: fired gunpowder,' *Proceedings of the Royal Society*, **153**. London.

Noble, A, 1890, *Address to the mechanical science section of the British association*. London.

Noble, A, 1894, 'Researches on explosives. Preliminary note,' *Proceedings of the Royal Society*, **56**. London.

Noble, A, 1900, *On some modern explosives, Royal Institution of Great Britain weekly evening meeting, Friday, March 23, 1900*. London.

Noble, A, 1907, *Fifty years of explosives, Royal Institution of Great Britain weekly evening meeting, Friday, January 18, 1907*. London.

Noble, A, 1909, 'A Sketch of the History of Propellants,' *Transactions of the Institution of Engineers and Ship-builders in Scotland* (reprint; volume and number not given). Glasgow.

Norman, A V B and Pottinger, D, 1985, *English weapons and warfare 449–1660*. New York.

Oman, C W C, 1924, *A history of the art of war in the middle ages*, 2 vols. New York.

Parker, G, 1988, *The military revolution: military innovation and the rise of the west, 1500–1800*. Cambridge.

Padfield, P, 1988, *Armada: A celebration of the four hundredth anniversary of the defeat of the Spanish Armada 1588–1988*. London.

Partington, J R, 1960, *A history of Greek fire and gunpowder*. Cambridge.

Pepper, S and Adams, N, 1986, *Firearms and fortifications: military architecture and siege warfare in sixteenth century Siena*. Chicago.

"Robins, Benjamin (1707–17561)", 1911, *Encyclopaedia Britannica*, 11th ed., **XXII**: 422. New York.

Rodman, T J, 1856, *Reports of experiments on the strength and other properties of metals for cannon*. Philadelphia.

Rodman, T J, 1861, *Reports of experiments on the properties of metals for cannon and the qualities of cannon powder, with an account of the fabrication and trial of a 15-inch gun*. Boston.

Tartaglia N, 1588 (original, Italian edition, 1546), *Three books of colloquies concerning the arte of shooting in great and small pieces of artillerie, variable randges, measure and weight of leaden yron and marble stone pellets, minerall saltpeeter, gunpowder of divers sorts and the cause of why some sortes of gunpowder are corned and some sortes of gunpowder are not corned*, C. Lucar, tr. & author of appendix. London.

Trebilcock, C, 1969, 'Spin-off in British economic history: armaments and industry, 1760–1914', *Economic history review*, **22**: 474 ff.

Tucker, S, 1989, *Arming the fleet: U.S. Navy ordnance in the muzzle-loading era*. Annapolis.

U.S. Army, 1964, *Fundamentals of ballistics: Special Text ST9–153*. U.S. Army Ordnance Center and School, Aberdeen Proving Ground, Maryland.

Williams, A R, 1975, 'The production of saltpetre in the middle ages', *AMBIX*, **XXII**.2: 125–33.

Williams, A R, 1974, "Some firing tests with simulated fifteenth century handguns,' *Journal of the Arms and Armour Society*, **VIII**, part 2, 114–120.

98

Fort Nelson The Royal Armouries Museum of Artillery

N HALL

Fort Nelson is one of five forts on Portsdown Hill to protect Portsmouth dockyard and the Naval installations from bombardment by rifled breech-loading guns.

Portsmouth has been fortified against landward attack almost since its foundation in the twelfth century. However, as the art of warfare progressed, it became necessary to update these defences, and, in time, Old Portsmouth, Gosport and Portsea were surrounded by ramparts and elaborate water defences. The land approaches had been defended by a fort since the reign of Henry VIII. This was rebuilt in 1746 during the War of the Austrian Succession. Further building work followed in 1756 during the Seven Years War, with the construction of the Hilsea Lines. The Napoleonic Wars caused further plans to be drawn up in 1812, but these were never put into effect.

The wars with Napoleon were to affect military and political thinking throughout the nineteenth century and France was always regarded as the most likely potential enemy. This feeling was given added momentum in 1848, when Louis Napoleon, nephew of Napoleon I, was elected President of the Republic. Concern became near panic in 1851, when he declared himself Emperor Napoleon III. There was a general fear that the new emperor would invade Britain to avenge Waterloo. This led to preparations to improve the defences of principal naval and military installations. Forts Gomer and Elson, built on the Gosport peninsular to protect Portsmouth's western approaches, date from this period.

War, however, did not break out between France and Britain. Indeed, the two countries formed an alliance with Turkey and took action against Russia in the Crimean War of 1854.

After peace with Russia was finalised, the British mistrust of France returned and when, in 1859, the French launched their steam-driven ironclad warship, the *Gloire*, there was a feeling that Britain's traditional command of the sea was being threatened by new developments in maritime technology. A ship of this type would be impervious to the artillery then in service. It might therefore be possible for France to land an army on the South Coast of England within the space of one night, without regard to the state of the wind and tide.

The sudden vulnerability of the British Isles to invasion became the key issue, and there was much talk of the 'steam bridge' and a need for improved defences. There was a loss of confidence in the Royal Navy, and with no large standing army (and no political will to provide for one) the solution appeared to be to provide fortifications around vulnerable areas. In order to facilitate this the Inspector General of Fortification, Sir John Fox Burgoyne, was asked to produce a report, and his assistant, William Drummond Jervois prepared a comprehensive plan for the defence of Portsmouth, which he submitted in December 1857.

Construction of the new Gosport forts and Hilsea Lines started in 1858. However, in 1859, William Armstrong's new rifled breech-loading gun made its appearance, and this weapon was to have an enormous impact on the design and positioning of fixed defences. In trials the rifled gun achieved a range of over 8000 yards with greatly improved accuracy. The existing and planned defences would therefore be unable to protect the dockyard from bombardment. If it was possible for an enemy to place guns on Portsdown Hill, he would be able to destroy the port, the town or anything else that he chose. It was clear, therefore, that the Portsdown Hill must be protected from the enemy. However, to defend a chalk ridge about seven miles long would require a whole army. This was not available, so the sensible alternative was to move the fixed defences out to the maximum range of the rifled gun.

Major Jervois recommended a continuous line of fortification, a ditch 'in the form of a railway cutting' along the entire hill, with caponiers every mile, projecting into the ditch to sweep it with enfilade fire. Detached works would be placed in front of the line with gun batteries to the rear. The line would be taken down to the shore on either flank, with a rampart of earth to obstruct the low points. Additional works would be situated on the western side of Fareham and at Newgate (Fort Fareham) to link up with the Gosport advanced line.

While Jervois was preparing his report, a committee of eminent naval and military officers met to consider the effect of the new gun. They recommended a line of some nine detached works along the length of the hill in place of Jervois' continuous line, which they felt could be breached and the line turned.

Government aid was needed to finance this vast project, and the government set up a royal commission to examine all the national defences in the light of the improvements in artillery. In 1860 the commission presented its findings to the government. The report concluded that to defend the whole coast was impracticable, and that new or improved defences should surround the principal naval and military establishments. For Portsmouth, the report recommended that a line of seven detached works should be built on Portsdown Hill. It suggested that five major works should be situated at Crookhorn (Fort Purbrook), Widley Mill (Fort Widley), the Fir Clump

(Fort Southwick), Nelson's monument (Fort Nelson) and above Wallington village (Fort Wallington), each about 2000–2500 yards apart. Between Crookhorn, Widley Mill and Fir Clump two smaller works were proposed. A continuous line would link all the works, the flanks of which would extend down to the shore. A line of forts would be built between Fareham and the Solent on the Gosport peninsula: Newgate (Fort Fareham), Roome and Lee Farm. The Hilsea lines were to be retained and completed as planned, but all the accessory work would not be carried out. In addition to this, five sea forts were to be built on the shoals of Spithead, and other batteries constructed at Southsea, Gosport and on the Isle of Wight. The estimated cost for the Portsmouth area was in the order of £4 million.

During 1860–61, the necessary land was acquired and the building of the works put out to tender. By mid-1861, the contracts were signed and work commenced. At the start of every financial year it was necessary to enact new legislation to raise money to pay for the works, and with each succeeding year this became more and more difficult. As a result the works were delayed, which in turn increased the costs still further. Despite considerable opposition, enough money was eventually found, however, and the works were largely completed by 1870. In this year, Napoleon III was goaded into war with Prussia, and any threat to Britain was fully removed when the French Emperor was taken prisoner at Sedan.

Much of the original enthusiasm for the forts had now waned, and the forts remained unarmed until the mid-1880s, when gradually they were equipped with armament, albeit on a reduced scale. In 1895 the whole concept of fixed defences changed. The greatly-increased ranges offered by modern armament meant that the forts no longer offered Portsmouth the protection for which they had been designed. A process of disarmament began, and by 1904 the forts had been stripped of their fixed artillery and were considered obsolete. In 1913, the forts were re-equipped for use as Royal Artillery depots and mobilisation stores. Married quarters were added, and stables were built for the horses. During the First World War, the forts became barracks for the Portsmouth garrison, at one time amounting to some 25,000 men.

Between the First and Second World Wars the forts were disused. In 1938, however, plans were prepared for turning Fort Nelson into a magazine for anti-aircraft gun ammunition. The parade was extended by cutting into the terreplein, a number of magazine buildings were erected, and the west gorge wall was pierced, to permit a one-way traffic flow for heavy lorries within the fort. The magazine was used a great deal during the war, but with the reduction in the armed forces in the 1950s, the fort was once again reduced to a state of care and maintenance. The Royal Navy stored electronic components in part of the fort until it was finally abandoned in the 1960s. After the fort had suffered years of neglect and vandalism, it was brought in 1979 by Hampshire County Council who then begun the task of restoration. In 1984 the work was advanced enough to allow the public to be admitted for the first time.

Already, the new Board of Trustees of the Royal Armouries were considering the proper housing and display needed for their large collection of artillery, which had long since outgrown the space available within the Tower's enceinte. Fort Nelson, though suffering from years of neglect, was unencumbered by preconceived ideas of re-use or existing commitments. Early visitor reactions were good, and its many gun positions and large barrack rooms seemed to offer a promising site for the Royal Armouries' first outstation. After negotiation Hampshire County Council enabled the Royal Armouries to take over the lease on Fort Nelson from 6 December 1988.

Prior to this date, guns from the Royal Armouries both from the Tower and on loan from elsewhere were being assembled at Fort Nelson. A land pattern, cast iron, 13 inch mortar had been placed in one of the original triple mortar batteries, later to be joined by a further two guns. Four cast iron, 32 pdr, SBBL guns on carriages and platforms were installed in the North Caponier. Artillery pieces which did not form part of the original armament of the Fort, have been displayed on the Parade and in the Barracks, where the fine bronze 12 pdr gun, *La Victoire*, captured from the prame 'La Ville de Lyon' in 1811 can be seen. At the conference, participants were able to watch members of the Portsdown Artillery Volunteer force perform gun drill on one of the 32 pdr SBBL guns in the North Caponier. This was the first such display at Fort Nelson and is a foretaste of activities to come. The Royal Armouries has recently acquired a 'light 110 pdr' (7 inch of 72 cwt) Armstrong RBL gun. This gun is particularly appropriate as an example of the kind of rifled gun that upset defence plans for Portsmouth and later became part of the armament of the fort. When installed in its Haxo casemate it should provide a fascinating exhibit, especially so when authentic gun drill is performed by members of the Portsdown Artillery Volunteers.